ROSENSTIEL SCHOOL OF MARINE AND ATMOSPHERIC SCIENCE,
UNIVERSITY OF MIAMI
4600 RICKENBACKER CAUSEWAY, MIAMI, FLORIDA 33149

STUDIES IN
TROPICAL
OCEANOGRAPHY
No. 14

This volume may be referred to as:

Stud. trop. Oceanogr. Miami 14: xii + 212 pp., 35 pls.

August, 1977

STUDIES IN TROPICAL OCEANOGRAPHY

STUDIES IN TROPICAL OCEANOGRAPHY NO. 14
ROSENSTIEL SCHOOL OF MARINE AND ATMOSPHERIC SCIENCE
UNIVERSITY OF MIAMI

Brachiopods from the Caribbean Sea and Adjacent Waters

By G. ARTHUR COOPER

UNIVERSITY OF MIAMI PRESS

Coral Gables, Florida

This volume may be referred to as:

Stud. trop. Oceanogr. Miami 14: xii + 212 pp., 35 pls.

August, 1977

Edited by

FREDERICK M. BAYER and GILBERT L. VOSS

Library of Congress Cataloguing in Publication Data.

Cooper, Gustav Arthur, 1902-

Brachiopods from the Caribbean Sea and adjacent waters.

(Studies in tropical oceanography; no. 14)

1. Brachiopoda—Caribbean Sea. 2. Brachiopoda—Mexico, Gulf of. 3. Brachiopoda—Atlantic Coast (United States). I. Title. II. Series.

QH91.A1S8 no. 14   [QL395.23]   551.4'6'00913s
ISBN 0-87024-277-6   [595'.32'0916365]   75-4757

FOREWORD

In 1963, a series of publications entitled *Studies in Tropical Oceanography* was established at the School of Marine and Atmospheric Science, University of Miami, to accommodate research reports too large for inclusion in regular periodicals. The first volume of this series was Dr. Donald P. de Sylva's "Systematics and Life History of the Great Barracuda, *Sphyraena barracuda* (Walbaum)," followed by 12 additional numbers covering a diversity of marine topics. As the fourteenth volume of *Studies in Tropical Oceanography* the editors now present an account of Recent brachiopods from the Gulf of Mexico and Caribbean Sea, based upon material from several sources including cruises by the research vessels of the Rosenstiel School of Marine and Atmospheric Science. This report, by Dr. G. Arthur Cooper of the Smithsonian Institution, is the first major work on the brachiopods of the Caribbean area in more than 50 years. Because of the low level of scientific interest in Caribbean brachiopods in recent years, together with the wide scope of the collections that have now become available for study, it was inevitable that many new taxa were discovered, changes in nomenclature made, and geographical distributions revised. This volume includes all the species at present known from the Gulf of Mexico and Caribbean Sea. It describes and illustrates the new taxa in detail and provides descriptive notes and comparative discussions of all species that heretofore were poorly known or likely to present difficulties in identification. It will become the new systematic standard for work on tropical western Atlantic brachiopods and will provide a firm foundation for further work on these neglected animals.

Dr. Cooper is one of the world's foremost authorities on brachiopods, both Recent and Fossil. His master work, "Permian Brachiopods of West Texas," is now appearing in the *Smithsonian Contributions to Paleobiology*. Having completed this enormous contribution to the knowledge of fossil brachiopods, Dr. Cooper is now devoting more time to the study of the Recent species, for which both biologists and paleontologists alike will be grateful.

The Editors

TABLE OF CONTENTS

ix

ILLUSTRATIONS

PLATES

TEXT FIGURES

TABLES

BRACHIOPODS FROM THE CARIBBEAN SEA AND ADJACENT WATERS

G. Arthur Cooper

Smithsonian Institution
(National Museum of Natural History)

ABSTRACT

Brachiopods occuring in the Caribbean region, Gulf of Mexico and off the Atlantic Coast as far north as New Jersey are recorded and described. Most of the specimens were taken on cruises of the University of Miami Research Vessels JOHN PILLSBURY and GERDA, the former operating in the Caribbean and the GERDA concentrating mostly on the straits of Florida. In addition to the magnificent collections of the University of Miami, specimens collected recently and in the past by U.S. Government vessels, especially the R/V OREGON and R/V SILVER BAY, through the kindness of Mr. Harvey Bullis, are included in the study. These combined collections are the greatest assemblage of Gulf and Caribbean brachiopods to be brought together, and they reveal hitherto unsuspected new genera and species. A few collections made by other vessels, off the Atlantic Coast from the Straits of Florida north to New Jersey, are included for the bearing they have on brachiopod distribution.

From the Caribbean and Straits of Florida 53 species are now recorded and assigned to 3 genera of the Inarticulata and 18 genera of Articulata. A single genus of the Rhynchonellida is recorded and all the other articulates are Terebratulida. Five genera are recorded from north of the Straits of Florida. These include *Abyssothyris* reported from the Atlantic for the first time. A new genus is described from the Bermuda Platform. From the Caribbean three new genera are recorded both with unusual structures, one with crural bases welded to the valve floor, the other with a doubly attached loop unlike any described previously.

Twelve genera are recorded from the Gulf of Mexico represented by 21 species. All of the genera and most of the species are also known from the Caribbean. Five of the species are new and only one of them occurs in the Caribbean as well as the Gulf. All three of the new genera from the Gulf also occur in the Caribbean.

Of the 32 new species described in this paper 17 belong to the new genus *Tichosina*, based on *T. floridensis*, n. sp. This is an externally generalized genus but it occurs throughout the Caribbean and Gulf of Mexico and is locally abundant. It is a vigorous genus with strong speciation. *Argyrotheca* is another common Caribbean genus with extreme speciation.

Most of the brachiopods recorded herein occur between tide level and the 1000 meter contour. Three are usually abyssal, and some species such as *Tichosina cubensis* (Pourtalès) *Dallina floridana* (Pourtalès) and *Argyrotheca barrettiana* (Davidson) have a wide depth tolerance. The last is the only species of *Argyrotheca* to have a very wide depth range.

1

INTRODUCTION

Brachiopods are rare in the modern seas and oceans, but they occur in all parts of the world and in all of the oceans. They are generally more frequently met in waters between the shallows and depths of 500 meters (Zezina, 1970: 5), but a great variety of rarer forms live in the depths between 500 meters and 6000 meters, the deepest that brachiopods have been encountered. Brachiopods are abundant in only a few places, such as around the Japanese Islands and in the Antarctic where Foster (1968: 160) reports them in a few local areas to be the most abundant part of the megafauna. Brachiopods have long been known from the Caribbean region, from which ten genera hitherto have been reported. Recent dredging by R/V JOHN ELLIOTT PILLSBURY and R/V GERDA from the University of Miami have taken brachiopods at 138 stations and have greatly enlarged our knowledge of these rare animals.

In 1963, I was invited by Dr. Frederick M. Bayer to study the brachiopod collections made by the PILLSBURY and GERDA on their numerous cruises. These vessels were used by the University of Miami in an extensive survey and study of the total tropical Atlantic to understand the community arrangement, composition and distribution of the benthic and midwater fauna including ecological and geographical aspects. Another goal is to understand the effect of ocean currents on the distribution of marine life between West Africa and South America. Special emphasis so far has been on the exploration of the Straits of Florida and the Caribbean. The former area was the special province of R/V GERDA. Here a grid of the Straits was made and trawling stations were occupied in as many of the squares of the grid as was possible during the program. The R/V PILLSBURY, on the other hand, has sampled areas off the west coast of Africa, and has closely surveyed the perimeter of the Caribbean Sea and the Antillean island arc, and has made transects across the basins.

These expeditions have taken brachiopods from the Straits of Florida where they have been known and are fairly common, and from the coasts of Yucatan, Honduras, Nicaragua, Panama, Colombia and Venezuela east to French Guiana. Specimens have been dredged from off Trinidad and along the west side of the Windward and Leeward Islands and north to the Bahamas. This is a vast and little known area and the diversity and novelty of many of the collections are ample testimony to their importance.

Not only have the University of Miami scientists been active in these regions but several vessels of the United States Fish Commission have made surveys in essentially the same places. As a matter of fact, a few of the localities of these vessels are almost precisely the same as some of these visited by the PILLSBURY. The U.S. R/V OREGON and SILVER BAY collected in the same region, the OREGON investigating as far east as the north coast of Brazil. Inasmuch as these collections often overlap and supplement the PILLSBURY and GERDA material, it seemed reasonable to combine the study of the OREGON and SILVER BAY collections with those of the PILLSBURY and GERDA. For example, GERDA took only four specimens of the new genus *Ecnomiosa* but OREGON collected two additional lots with a more abundant representation. In addition, a few collections made in the last century by the BLAKE, FISH HAWK and ALBATROSS, which cruised in the Caribbean, have been included together with a few isolated specimens of great interest from off the coasts of Georgia, South Carolina and New Jersey. The presence of brachiopods in these areas has not been generally known and some interesting new forms, such as *Abyssothyris,* are of great importance in brachiopod distribution and evolution. This report also includes a new genus collected by Dr. Heinz Lowenstam, California Institute of Technology, from the Bermuda Platform. In all 208 locations of brachiopods are recorded which include a generous number of interesting new species and three new genera. This paper endeavors to record all other brachiopods recorded from the Caribbean, the Straits of Florida, the coastal region north of the Straits to New England, and the Gulf of Mexico.

As might be expected from such an abundance of riches, wider distribution of old species has been established and new forms have been discovered. Twenty-one genera of which three are new were taken during the activity of the vessels mentioned above from the Caribbean and east coast of the United States. From the same region, 54 species are now reported of which 27 are new. Some may wonder at the abundance of new species but it should be pointed out that the brachiopods of this region have not been subjected to any revision since the 1920's and that the collections include specimens from many areas not hitherto visited. Some Caribbean species of generalized form, such as "*Terebratula*" *cubensis* Pourtalès, have been widely and indiscriminately identified. The result was a wider distribution than it really has and a name that included a variety of distinct species. The loops of the Terebratulidae, to which "*T.*" *cubensis* belongs, have never been closely analyzed. When this is done, more genera or subgenera can be expected. When presented with an extensive collection from little-known regions, any investigator is confronted with the problem of species recognition. It is unlikely that his collection of individual species will be large enough to embrace all their variations or local aberrations. It is best, therefore, to name reasonably good material where it appears different because it will take years before collections will be large enough to know all about variation and geographic distribution.

All species of brachiopods from the Gulf of Mexico known to me are recorded with notes and descriptions of poorly known species. Names erroneously listed from the Gulf are corrected. Twelve genera and 21 species have been identified, indicating a fairly generous representation in the Gulf of this intriguing group of invertebrate animals. Three of the genera are new, *Tichosina, Stenosarina* and *Ecnomiosa.* Six new species are described: *Tichosina floridensis, T. ovata, Stenosarina angustata, S. oregonae, Ecnomiosa gerda,* and *Argyrotheca hewatti.* Each of the new genera and *T. floridensis* also occur in the Caribbean.

All of the genera occuring in the Gulf of Mexico are also known in the Caribbean Sea. *Thecidellina* and *Lacazella* have not yet been taken from the Gulf except for the report of the latter in the older literature, a report not yet confirmed by modern collecting.

3

ACKNOWLEDGMENTS

I am indebted to Dr. Frederick M. Bayer for making this study possible and to Dr. Gilbert L. Voss both of the University of Miami, Florida, for releasing the types and illustrated specimens to the National Museum. To Mr. Harvey Bullis go thanks for the fine collection of Caribbean brachiopods given to the Museum and the opportunity to study this valuable material. Thanks are given Dr. Heinz Lowenstam, California Institute of Technology, for his gift of specimens on which the new genus *Notozyga* is based. I also express my gratefulness to Dr. Jeremy Jackson of Johns Hopkins University for his gift of specimens of *Argyrotheca, Thecidellina* and *Lacazella* from Jamaica where these genera are proving to be abundant in grottos and on the undersides of corals.

It is a pleasure to acknowledge the help of several collectors and scientists who have contributed to the collection from the Gulf of Mexico. The specimens presented by Mr. Harvey Bullis constitute the largest part of the Gulf of Mexico material. Dr. Willis E. Pequegnat, Department of Oceanography, Texas A. and M. University, presented specimens of the deep water *Chlidonophora* and two rare specimens of *Platidia anomioides* (Scacchi and Philippi). Dr. Willis G. Hewatt, Texas Christian University, presented the specimens of *Argyrotheca* that are to bear his name. A fine lot of specimens was presented by Robert W. Cooper, of the Cooper Museum of Marine Invertebrates, Peoria, Illinois. To all of these contributors go the thanks of the Smithsonian Institution and myself.

I am grateful to Drs. Richard E. Grant and J. Thomas Dutro, Jr., for reading the manuscript and making valuable suggestions. I acknowledge with thanks the excellent drawings of the muscles of *Megerlia* and *Argyrotheca rubrocostata,* n. sp., made by Mr. Lawrence B. Isham, Visual Information Specialist, Department of Paleobiology, National Museum of Natural History.

PROBLEMS IN THE STUDY OF BRACHIOPODS

Although brachiopods have been known and studied for two centuries, many details of their biology are still to be described. Their feeding habits and what they feed on as well as other physiological details are now for the first time being investigated. The embryology of only five genera is known in detail. Furthermore, what constitute generic and specific characters are not agreed on by all workers. Brachiopod ecology is now coming under modern scrutiny (Paine, 1963; Foster, 1968; Logan, 1974).

Brachiopods have been studied, classified and described mostly by paleontologists. Inasmuch as brachiopods were among the earliest animals to appear in the geological column and, through the Paleozoic, were perhaps the most abundant animals of the marine megafauna, the paleontologist has had more to say about them than the zoologist. The paleontologist is also responsible for the present classification of brachiopods, Recent as well as fossil (Williams and Rowell, 1965 *in* Williams, A. 1965).

Species are based largely on shell character. This is true for the Recent as well as fossil forms. The brachiopod shell has but a limited number of characters; combinations of these are used to recognize genera and species. In considering the paleontologist's species one must remember that he is dealing with a time element as well as geographic distribution in the creation of species and genera. He needs the minute characters to help him detect trends of evolution.

GENERIC AND SPECIFIC CHARACTERS IN RECENT BRACHIOPODS

In a recent essay on the generic characters of brachiopods Cooper (1970) noted that a large number of fossil brachiopod genera are based on exterior characters consisting of ornament, costellae, costae or plicae, spines, rugae, lamellae, the folding of the anterior commissure and, occasionally, an unusual shape. The family characters of most of the fossil brachiopods are to be found on the interior: patterns of cardinalia, muscle scars, loops and other details of the inside. To some extent, use of exteriors as generic characters is employed in the modern brachiopods. For example, the anterior commissure of *Abyssothyris* is folded toward the ventral side, an unusual type of folding; *Waltonia* is an externally smooth genus whose interior is exactly like costate *Magasella*; *Rhytirhynchia* is an anteriorly plicated *Basiliola*; *Argyrotheca* is strongly plicated, but *Cistellarcula* has subdued costellae. Actually, establishment of a genus involves a combination of characters, external and internal, to differentiate it from known genera. Seldom does a single character identify a new genus. *Erymnia*, described herein, however, is an unusual example of a unique feature being the sole basis for establishment of a genus.

GENERIC CHARACTERS

Features of the exterior that have been used in genus-making are: ornament, beak and hinge characters, and anterior commissure. Few of the Caribbean brachiopods are highly ornamented and none of the genera is based primarily on ornament. Examples outside of the Caribbean are found in New Zealand where *Terebratalia* and *Diestothyris* have identical loop characters but the former is costate whereas the latter is smooth.

Beak characters are important in generic separation, but some have been carried to extreme. Most members of the Cancellothyridacea, of which *Terebratulina, Eucalathis* and *Chlidonophora* of the Caribbean are examples, have a large foramen and the deltidial plates usually are vestigial. The hinge may be wide or narrow but usually it has small ears that are quite helpful in identifying a member of this family. An aberration of *Terebratulina,* with conjunct deltidial plates and a somewhat restricted foramen, has been named *Cancellothyris* even though all other details are like *Terebratulina.* *Aerothyris* and *Magellania* with identical loops are separated only on the basis of the condition of the deltidial plates: disjunct in the former, conjunct in *Magellania.*

The folding of the anterior commissure is used as a generic character in a number of genera. *Abyssothyris* is distinguished externally from most other terebratulinids by its sulcate anterior commissure, i.e., the folding is toward the ventral side, the opposite of the folding tendency in most of the Caribbean species here referred to *Tichosina.* It is interesting to note that

5

sulcate folding occurs in two other brachiopods that often occur with *Abyssothyris* in the great deeps (Cooper, 1972). However, at present it seems impossible to link sulcation with depth because a number of Recent shallow water brachiopods are sulcate. An example is *Waltonia* that lives in the tide zone in Wellington Harbor, New Zealand.

Internal generic characters are commonly found in the loop and cardinalia. As examples of the latter, *Macandrevia* and *Dallina,* both described herein and both with adult unsupported loops, may be cited. *Dallina* is distinctive not only for its complicated anterior commissure but also for the union of the inner hinge plates with a prominent median septum to form a small apical chamber that lodges the pedicle muscles. *Macandrevia* has no median septum, and the inner hinge plates descend to the floor of the valve and attach there to form an elongate chamber. Extensions from the anterior ends of these plates stretch for some distance along the valve floor. The metamorphosis of the loop of these two genera is almost identical (see below) but the cardinalia patterns are completely different.

Study of the interiors illustrated herein will show other distinctive cardinalia patterns that are rather more familial than generic characters. For example, the cardinalia of *Terebratulina* are essentially the same as those of *Eucalathis* and *Chlidonophora,* which have different loops and different exterior characters.

The metamorphosis of the loop of the short-looped terebratulinids is not elaborate. Although the brevity of the loop is a major taxonomic character, various aberrations of the loop have familial and generic value. The terebratulinid type of loop serves as a good example. In the Cancellothyridacea, of which *Terebratulina* is an example, the crural processes form a ring, but in *Eucalathis* and *Chlidonophora,* now in separate subfamilies, the same elements of the loop are present but the crural processes never form a ring and the transverse ribbon is extended in a sharp curve in an anterodorsal direction.

The genus *Tichosina* is created in this paper for certain short-looped forms in which the crural base is continued posteriorly along the inner edge of the outer hinge plate as a fairly high wall. An example is *T. cubensis* (Pourtalès) which had hitherto been placed in the genus *Dallithyris* (Muir-Wood, 1959), a genus having a shape almost identical to that of *T. cubensis.* The loops, however, are very different because that of the type-species of *Dallithyris* (*murrayi* Muir-Wood) does not have the crural bases extended along the outer hinge plate. It is most like the loop of *Gryphus vitreus* (Born) of the Mediterranean, which has flat and unbordered outer hinge plates. Small but persistent aberrations of the loop offer the best possibilities for separating the various stocks of these prolific but difficult-to-classify brachiopods. The genus *Erymnia* proposed herein affords an unexpected example of a trend in loop development. The loop of *Erymnia* is like that of *Tichosina* in most of its details but it is welded to the valve floor by struts built along the dorsad edge of the crural base. So far as is known, this is the first example of a terebratulid with the loop welded to the valve floor in such a manner.

The long-looped brachiopods such as *Dallina* are instructive in the taxonomy of brachiopods. The family Dallinidae is defined as characterized by a certain metamorphosis of the loop. In this development a triangular pillar is formed on the floor of the early shell to which the crura or descending lamellae are united. A ring buds from the crest of the pillar, now a septum, and expands into a hood-like structure very much like that in the adult brachiopod *Campages.* As growth and expansion continue the brachiopod resorbs part of the elements so that the next stage suggests that seen in adult *Terebratalia,* in which the descending branches of the loop are attached to the median septum but the ascending branches are free. The final stage is resorption of the lateral branches attaching the loop to the septum and part or all of the septum, resulting in a completely free loop supported only by the hinge plates. This is the ultimate in the development of all the long-looped brachiopods. The development of loop just outlined is similar to that of the Terebratellinae, differing in minor details but resulting in an unsupported loop, an example of which is *Magellania.* When it is demonstrable that an arrested loop is in the adult condition, genera have been made to define that condition. Examples of this are the two genera recorded above: *Campages* and *Terebratalia.*

A remarkable example of the development of a long loop is that of *Economiosa* described herein. The adult loop of this genus is not in the ultimate free condition because it is attached posteriorly to a short median septum. Unfortunately, we do not have its full developmental history. At an early stage the loop is in the condition of *Campages,* with a large hood and the descending lamellae attached to the median septum. Later growth to a large size sees the partial or complete elimination of the lateral supports to the descending lamellae but the posterior ring is not eliminated. The complicated nature of this loop is clearly generic because of the ring-like attachment retained by the adult. This necessitates a family status and further development within the family is possible by resorption of the posterior ring and the complete freeing of the loop (see plates 27 and 28).

SPECIFIC CHARACTERS

The specific characters of most of the brachiopods described herein are based largely on exterior features: shape or outline, lateral and anterior profiles, convexity of valves, lateral anterior commissure aberrations and ornament if any. In some species the shape of the loop proves to be a supplementary character. Use of such characters involves considerable personal bias, but this bias tempered with experience and care in selecting adequate specimens for description will insure a minimum of splitting.

The shell of terebratulid brachiopods is distinguished by perforations or small holes, called punctae, which are occupied by caeca of the mantle. Attempt has been made to use the number of punctae in a square millimeter as help in distinguishing species. These are counted on the slope anterior to the ventral umbonal region. So far this has not proved very satisfactory. The count is not always the same in parts of the same specimen. Counts are given herein in the hope that they will ultimately prove useful.

TABLE 1
Table showing specific characters of the species of *Tichosina* described herein.

Species	Size	Shape	Foramen	Anterior commissure	Loop	outer hinge plate	Transverse band	Flexure transverse band	Crura	Punctae per square mm
abrupta Cooper, n. sp.	s	No	L	FU	wp	ld	B	G	sh	?
bahamienesis Cooper, n. sp.	s	o	s	R	Np	shc	B	S	l	116
bartletti (Dall)	L	Eo	s	SU	Np	shd	B	G	l	77-119
bartschi Cooper	A	No	s	R	FT	ld	B	NS	l	64
bullisi Cooper, n. sp.	L	ro	L	FU	FT	lc	B	B	sh	84
cubensis (Pourtalès)	L	T-P	L	FU	Np	mld	B	NS	sh	77-95
dubia Cooper n. sp.	A	rp	L	GU	wFT	ld	B	G	sh	99
elongata Cooper, n. sp.	A	No	L	R	Np	?	B	S	sh	75
erecta Cooper, n. sp.	A	o	s	R	Np	shfl	B	S	sh	65
expansa Cooper, n. sp.	L	ro	L	BU	Np	Bflc	B	Br	sh	100
labiata Cooper, n. sp.	A	To	L	BFU	Fw	ld	B	G	sh	102
marticiensis (Dall)	A	ro	s	R	Np	nsh	B	Br	sh	158
obesa Cooper, n. sp.	s	Eo	s	R-SU	wp	dc	B	nG	sh	135
ovata Cooper n. sp.	L	Eo	mL	m-sU	Np	dc	B	S	sh	70-100
pillsburyae Cooper, n. sp.	L	o	s	SU	p	shd	B	m	ml	76-95
plicata Cooper, n. sp.	A	o	L	SU	T	lc	B	m	vsh	132

rara Cooper, n. sp.	A	ro	s	BU	wp	ld	mB	S	1	75-78
rotundovata Cooper, n. sp.	A	Eo	s	R	Np	Lfl	mB	B	sh	82-100
solida Cooper, n. sp.	L	Eo	L	BGU	Np	wc	B	NS	1	78-84
subtriangulata Cooper, n. sp.	A	ST	s	R-BU	Np	nd	vB	N	sh	102
truncata Cooper, n. sp.	A	Eo-ET	s	R	Np	Bc	B	N	1	63-72
sp.1	A	ro	s	nU	Np	shc	B	NS	1	89
2	A	Eo	L	FU	p	?	B	B	?	?
3	s	No	L	SnU	Np	ld	?	?	1	?
4	L	Po	L	GU	?	?	?	?	?	?
5	L	Eo	L	GU	?	?	?	?	?	?
6	A	To	L	BU	w	Nshc	?	?	sh	?
7	A	Eo	L	BU	T	Nd	B	N	sh	140
8	L	Eo	L	BU	p	ld	B	B	sh	76-79
9	A	Eo	s	U	Np	ld	B	mB	sh	?

Explanation of symbols

A = average size (20-30 mm)
B = broad or broadly
c = concave
d = deep or deeply
E = elongate
F = faint or faintly

fl = flat or flatly
G = gentle or gently
L = large
l = long
m = moderate or moderately
N = narrow

o = oval
P = pentagonal
p = parallel-sided
R = rectimarginate
r = round or roundly
S = strong

s = small
sh = short
ST = subtriangular
T = triangular
U = uniplicate
v = very
w = wide

According to Zezina (1970: 5) the majority of the brachiopod species live at depths from tide zone to 500 meters, the outer edge of the continental shelf. This is true of the brachiopods treated herein. Ten species are restricted to shallow water (0-183 meters = approximately the upper edge of the continental shelf) but 20 are restricted to waters from 0-500 meters. Ten are restricted to still deeper water. These are genera that have not been found in water shallower than 500 meters: *Abyssothyris, Chlidonophora* and *Pelagodiscus*. Several Caribbean genera have a fairly wide depth tolerance. *Argyrotheca barrettiana* (Davidson) and *A.* sp. 3 are the only species of *Argyrotheca* found in water deeper than 183 meters. *Dallina floridana* (Pourtales) ranges from 100 down to 724 meters but is commonest between 100 and 250 meters. *Tichosina cubensis* (Pourtalès) has a wide depth range, from 183 meters down to 962 meters, but the majority of the specimens in the collection come from about 250 meters. *Platidia anomioides* (Scacchi and Philippi) ranges from 33 to 963 meters and *Terebratulina cailleti* Crosse extends from 33 meters to 963 meters. These long ranges of the commonest species are a warning to paleontologists not to rely too heavily on depth ranges of modern brachiopod genera in speculations on depth of fossil faunas and sediments (see Table 2).

RELATION TO OTHER RECENT FAUNAS

Of the brachiopod associations in the world, the Caribbean brachiopod fauna is closest to that of the Mediterranean. Eight of the 21 Caribbean genera are present in the Mediterranean: *Lacazella, Dallina, Platidia, Terebratulina, Argyrotheca, Megerlia, Eucalathis* and *Crania*. Six of the 21 genera are also present in the eastern Atlantic: *Macandrevia, Dallina, Cryptopora, Crania, Terebratulina, Eucalathis* and *Dyscolia*. Of Mediterranean elements *Crania, Eucalathis, Dyscolia* and *Megerlia* are all very rare. Although *Dyscolia* was first discovered in the Caribbean, it has not been seen since then until the recent find by the PILLSBURY reported herein. *Platidia* and *Terebratulina* are fairly common in both regions, particularly *Terebratulina*. *Argyrotheca* is common in the Mediterranean but its abundance in the Caribbean is unmatched elsewhere. *Cryptopora*, the only rhynchonellid known in the Recent fauna of the Caribbean, is world-wide in occurrence. This is true also of the Inarticulate *Pelagodiscus*. *Glottidia* is restricted to the southern coasts of North America, west coast of Central America and is known from Puerto Rico. Its report from Martinique is doubtful (Dall, 1920: 269). The new genera described herein do not indicate close relationships to any other realm of the brachiopods.

The Caribbean fauna may be thought of as dominated by *Tichosina* (hitherto called *Gryphus* or *Dallithyris*, neither of which now occurs in the Caribbean). The next most abundant element is *Argyrotheca*. *Thecidellina*, which is fairly common in the Caribbean, is unknown in the Mediterranean although it occurs in the Pacific and Indian Oceans, fairly common in the former but very rare in the latter. The Caribbean fauna is just as characteristic of this region as is the terebratelloid dominated area of New Zealand and the varied and distinctive faunas about Japan and the Antarctic.

It should be noted that *Macandrevia,* an element of the Mediterranean and Atlantic faunas, fossil and Recent, approaches the Caribbean as far as New Jersey and the west coast of Africa, but in its Tertiary migration to the west coast of the Americas, it apparently shunned the Caribbean. If the migration was from north Atlantic or Mediterranean waters through the Panama region to the west coast of Panama and South America to the Antarctic, some should have lingered in the Caribbean but none has yet been found in these modern waters or in Tertiary sediments. *Macandrevia* is a deep- or cold-water loving brachiopod and may ultimately be found in the deeper waters of the Caribbean.

ORIGIN OF THE CARIBBEAN FAUNA

The end of the Cretaceous was a critical time for the brachiopods. In that period the short-looped terebratulids were abundant and the rhynchonellids were common. After the end of the Cretaceous only three Articulate genera are known to have survived into the Eocene: *Terebratulina, Argyrotheca* and *Megathiris.* The first two survived into the Eocene in the Caribbean and east and Gulf Coast regions of the United States. With them appeared the Mediterranean *Gryphus* in the Eocene of Cuba but that genus is not now known in Caribbean waters, nor did it survive to the Miocene in the Caribbean region. *Thecidellina* and *Lacazella* are the only other genera present in the Caribbean today that appeared in the Eocene of the east and Gulf coasts of the United States and the West Indies (Cuba). *Platidia* appears in the Oligocene and Miocene of Cuba. In the Eocene of Cuba *Phragmothyris,* an extinct form related to *Argyrotheca* (Cooper, 1955) and the extinct Mediterranean rhynchonellid *Erymnaria,* enjoyed a short existence (Cooper, 1959: 64). In the Eocene of the east coast of the United States, two terebratulid genera did not survive that period: *Oleneothyris* and another yet to be described. But specimens collected from the Eocene of Cuba have all the characters of *Tichosina,* a genus that survived to the present. *Cryptopora* is another brachiopod appearing in the Miocene or later Tertiary of Cuba. *Dallina, Macandrevia, Megathiris, Megerlia,* and *Crania,* all Mediterranean Tertiary genera, have not yet been taken from the Tertiary of the Caribbean region. A homeomorph of *Dallina,* but having an entirely different interior morphology, occurs in the Eocene of Cuba but now is extinct. The origin and composition of the Eocene brachiopod fauna is not now well known but its importance to the origin of the modern brachiopod faunas cannot be overemphasized.

RELATION OF THE GULF BRACHIOPODS TO THOSE OF THE CARIBBEAN

Cooper (1954) recorded 12 species from the Gulf of Mexico. These names were listed from specimens in the National Museum of Natural History and at Harvard University. The latter collections were taken by Pourtalès, who was one of the early oceanographic explorers in the Gulf of Mexico. The Cooper list embraces the names known at that time, but it must be revised in the light of present and newly acquired information. The species listed as *Cryptopora gnomon* Jeffreys is now identified as *Cryptopora rectimarginata* Cooper, a species fairly common in parts of the Straits of Florida. The

11

TABLE 2
Depth in meters

	0-100	101-500	501-1000	1001-2000	2001-3000	3001-4000
Abyssothyris atlantica n. sp.					●	
A. ? parva n. sp.						
Argyrotheca barrettiana (Davidson)	●		●	●		
A. bermudana Dall	●					
A. crassa n. sp.	●					
A. hewatti n. sp.						
A. johnsoni Cooper	●					
A. lutea (Dall)	●					
A. rubrocostata n. sp.	●					
A. rubrotincta (Dall)	● ●					
A. schrammi (Crosse and Fischer)	●					
A. woodwardiana (Davidson)		●				
A. sp. 1	●					
A. sp. 2	●					
A. sp. 3						
Chilidonophora incerta (Davidson)		●	●	●		
Crania pourtalesi Dall		●				
Cryptopora rectimarginata Cooper		●				
Dallina floridana (Pourtalès)		●				
Discradisca antillarum (D'Orbigny)						

	0-100	101-500	501-1000	1001-2000	2001-3000	3001-4000

Dyscolia wyvillei (Davidson)

Ecnomiosa gerda n. sp.

Erymnia angusta n. sp.

E. muralifera n. sp.

Eucalathis cubensis n. sp.

E. floridensis n. sp.

Glottidia pyramidata (Stimpson)

Lacazella caribbeanensis n. sp.

Macandrevia novangliae Dall

Megathiris detruncata (Gmelin)

Megerlia echinata (Fischer and Oehlert)

Notozyga lowenstami n. sp.

Pelagodiscus atlanticus (W. King)

Platidia anomioides (Scacchi & Philippi)

P. clepsydra Cooper

P. davidsoni (Deslongchamps)

Stenosarina angustata n. sp.

S. nitens n. sp.

S. oregonae n. sp.

S. parva n. sp.

Terebratulina cailleti Crosse

T. latifrons Dall

Thecidellina barretti (Davidson)

TABLE 2 (Continued)
Depth in meters

	0-100	101-500	501-1000	1001-2000	2001-3000	3001-4000
Tichosina abrupta n. sp.		▌				
T. bahamiensis n. sp.		▌				
T.? bartletti (Dall)	▌	▌	▌			
T. bartschi Cooper		▌	▌			
T. bullisi n. sp.		▌				
T. cubensis (Pourtalès)		▌	▌			
T. dubia n. sp.		▌				
T. elongata, n. sp.			▌			
T. erecta, n. sp.		▌				
T. expansa n. sp.			▌			
T. floridensis, n. sp.		▌				
T. labiata, n. sp.		▌				
T. martinicensis (Dall)		▌				
T. obesa, n. sp.	▌	▌	▌			
T. ovata, n. sp.		▌				
T. pillsburyi, n. sp.		▌				
T. plicata, n. sp.		▌				
T. rotundovata, n. sp.			▌			
T. solida, n. sp.		▌				
T. subtriangulata, n. sp.		▌	▌			
T. truncata, n. sp.		▌				

specimen from the BLAKE collection identified as *Gryphus bartschi* Cooper is clearly not that species because it has a small foramen and entirely different shape. It belongs to *Tichosina floridensis* n. sp., described herein. The specimens referred to *Argyrotheca schrammi* (Crosse and Fisher) are best identified as *A. rubrotincta* (Dall), the type specimen of which comes from the Dry Tortugas.

All of the other species, *Glottidia pyramidata* (Stimpson), *Chlidonophora incerta* (Davidson), *Terebratulina cailleti* Crosse, *Gryphus* (now *Tichosina*) *cubensis* (Pourtalès), *Argyrotheca barrettiana* (Davidson), *A. lutea* (Dall) and *Dallina floridana* (Pourtalès) occur in the Caribbean. Too few specimens have been found of most of these species to know whether or not subspecific distinctions exist in the two water bodies. *Chlidonophora* is restricted to deep water in both the Caribbean and the Gulf. The species of *Argyrotheca* of the Gulf are generally confined to fairly shallow water. Some are in deep water in the Caribbean. *Terebratulina cailleti* is the widest spread of the species in both geographic and bathymetric range.

Tichosina cubensis (Pourtalès) from the Gulf presents no differences from specimens taken from the Caribbean. Those of the Caribbean range in depth from 88-400 fathoms (=161-732 meters)(Dall, 1920: 315). One specimen however, is recorded by Dall as coming from 2690 fathoms (=5003 meters) from "off Cuba." This specimen on examination proves to be an immature brachiopod with an outline suggestive of *T. cubensis* but rather narrower and more elongate. Immature *T. cubensis* is subcircular and not elongated like the specimen in question. That specimen (USNM 110855) is clearly not the young of *T. cubensis* and definitely belongs to some other species not now known or recognizable. In the Gulf a very large and well formed adult *T. cubensis* (USNM 550755) was taken by the OREGON (Station 1189) at 1700 fathoms (=3111 meters). This is the deepest positive record for the species in either body of water.

Dallina floridana (Pourtalès) presents no complications as regards resemblance or bathymetry in the two water bodies. This species in the Caribbean occurs from 90-270 fathoms (=165-491 meters). The few specimens from the Gulf range from 295 to 457 meters. One immature specimen (USNM 432772) assigned to this species comes from 35 fathoms (=64 meters).

ORIGIN OF THE BRACHIOPOD FAUNA OF THE GULF OF MEXICO

Most of the genera occurring in the Gulf of Mexico have antecedents in the Tertiary of the Atlantic coast of the United States and a few in the Tertiary of the Gulf coast. All but one of the Inarticulate genera of the Gulf are represented in the Tertiary of the Atlantic coast but only *Crania* has been found in the Gulf coast Tertiary. *Glottidia inexpectans* Olsson occurs in the Miocene (Yorktown Formation) of Virginia. *Pelagodiscus,* a deep sea brachiopod, has not yet been recognized in the fossil state in Tertiary sediments of the United States or the Caribbean region. Deep sea sediments are not usual in the geological record. Its authentication as a fossil will have to await positive identification of deep sea deposits. *Crania* occurs in the Eocene of the Atlantic and Gulf coasts but it has not yet been reported from

15

later Tertiary sediments. The Articulate genera are better represented as fossils in the Tertiary, but they are rare in these deposits along the Gulf coast.

Six of the Articulate genera reported herein from the Gulf of Mexico occur in the Atlantic and Gulf coasts or in the Tertiary of the Caribbean. *Cryptopora* is known from the Eocene of Alabama and the Oligocene-Miocene of the Caribbean. *Terebratulina* occurs in the Eocene and later Tertiary sediments of each region. It was prolific as a fossil and is very abundant in modern seas. *Tichosina* is present in the Oligocene of Cuba and *Stenosarina* seems to have antecedents as far back as the Cretaceous. It is present in the Eocene of Cuba. *Platidia* occurs in the Eocene and the Pliocene of Cuba and the Eocene of the Gulf coast. *Argyrotheca*, with its roots back into the Mesozoic, is common in the Eocene of the Gulf and Atlantic coast Tertiary and is also common in the Eocene and Miocene of Cuba. Thus six of the genera of the Gulf have representatives in the Tertiary.

Several genera have not yet been found as fossils in Caribbean or Gulf Tertiary sediments. *Dallina* is not known as a fossil in these regions, nor is *Chlidonophora*. Like *Pelagodiscus*, that genus is a deep sea form and it is unlikely to be found as a fossil. Its report from the Cretaceous is certainly an error. *Ecnomiosa* is unknown as a fossil.

While mentioning those genera known as fossils, it is interesting to note that a number of the genera that appeared as fossils in the Eocene are now extinct. These are: the modern Mediterranean genus *Gryphus* from the Eocene of Cuba; *Oleneothyris* from the Eocene of the Atlantic coast; and *Phragmothyris*, a genus related to *Argyrotheca*, found in the Eocene of Cuba. Two rhynchonellid genera flourished in the Eocene: *Erymnaria* from the Eocene of Cuba and southern Europe, and *Probolarina* from the Eocene of the Atlantic coast.

Two important genera that are fairly common in the Caribbean are not yet certainly known from the Gulf of Mexico. These are the cemented genera *Thecidellina* and *Lacazella*. Davidson (1887: 158) reported *Lacazella mediterranea* (Risso) from the Gulf of Mexico in collections made by the U.S. Coast Survey steamer BLAKE from 293 meters. No specimens of this species are now in the BLAKE collections belonging to the National Museum of Natural History. Both of these occur in the Eocene of the Gulf coast and they are present in the Eocene and Miocene of Cuba. They commonly inhabit reefs or coralline structures and should be looked for on such limy accumulations.

Although most of the Caribbean fauna has been resident since Tertiary times, a large element of the Gulf fauna arrived rather late in geologic time.

DREDGING STATIONS IN THE ATLANTIC, CARIBBEAN AND STRAITS OF FLORIDA

The dredging stations and their listed brachiopods include those of the research vessels belonging to the United States government: ALBATROSS, BLAKE, COMBAT, FISH HAWK, OREGON and SILVER BAY. Dredging stations in the Gulf of Mexico include all of the above government vessels except the COMBAT, the OREGON and SILVER BAY accounting for most of the material.

ALBATROSS

2035   Latitude 39° 16′ N, longitude 70° 02′ 37″ W at 1362 fathoms (= 2492 meters), east of Atlantic City, New Jersey, figure 1.
Abyssothyris? sp. 1
Macandrevia novangliae Dall

2208   Latitude 39° 33′ 00″ N, longitude 71° 16′ 15″ W at 1178 fathoms (= 2156 meters), south of Block Island, east of New Jersey, figure 1.
Macandrevia novangliae Dall

2220   Latitude 39° 43′ 30″ N, longitude 69° 23′ 00″ W at 1054 fathoms (= 1929 meters), off Nantucket Shoals, New England, figure 1.
Macandrevia novangliae Dall

2152   2.5 miles northwest of Havana Light at 387 fathoms (= 708 meters), not on map.
Tichosina subtriangulata, n. sp.

2343   Latitude 23° 11′ 35″ N, longitude 82° 19′ 25″ W at 279 fathoms (= 511 meters), northeast of Havana, Cuba, figure 4, number 66.
Tichosina subtriangulata, n. sp.

2535   Latitude 40° 03′ N, longitude 67° 27′ 15″ W, southeast of Georges Bank at 1149 fathoms (= 2103 meters). Not on map.
Macandrevia novangliae Dall

2655   Latitude 27° 22′ N, longitude 78° 07′ 30″ W at 338 fathoms (= 619 meters), north side of Grand Bahama Island, figure 4, number 67.
Tichosina bahamiensis, n. sp.

2682   Latitude 39° 38′ N, longitude 70° 22′ W at 1004 fathoms (= 1837 meters) off Marthas Vineyard, Massachusetts, figure 1.
Macandrevia novangliae Dall.

2706   Latitude 41° 28′ 30″ N, longitude 65° 35′ 30″ W, at 1188 fathoms (= 2173 meters), 240 miles east of Nantucket, Massachusetts, figure 1.
Macandrevia novangliae Dall

17

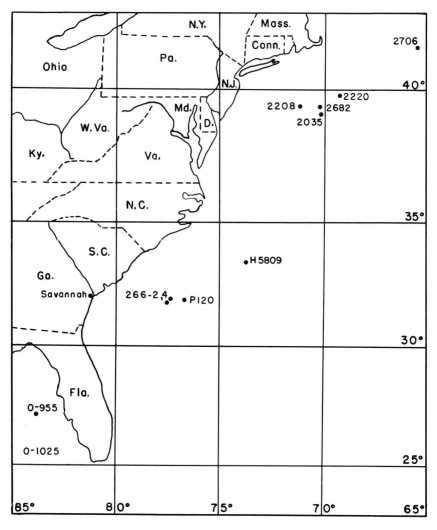

FIGURE 1. Map of east coast of United States, showing location of HYDRO, ATLANTIS, some ALBATROSS and one PILLSBURY station. ALBATROSS STATIONS: 2035, 2208, 2220, 2682, and 2706. ATLANTIS Stations: 266-2, 266-4. HYDRO Station: H5809. PILLSBURY Station: P120.

266-2   Latitude 31° 58′ N, longitude 77° 18′ W, at 445 fathoms (= 734 meters) east of Savannah, Georgia, figure 1.
Abyssothyris? parva, n. sp.

266-4   Latitude 31° 56′ N, longitude 77° 26′ W at 420 fathoms (= 769 meters), off Fernandina Beach, Florida, figure 1.
Abyssothyris? parva, n. sp.

BLAKE STATIONS

16   Latitude 23° 11′ N, longitude 82° 23′ W at 785 fathoms (= 1437 meters), off Havana, Cuba, Figure 2; figure 4; number 68.
Chlidonophora incerta (Davidson)
Eucalathis cubensis n. sp.
Platidia anomioides (Scacchi and Philippi)

147   Latitude 17° 19′ 27″ N, longitude 62° 50′ 30″ W, at 250 fathoms (= 458 meters), off west side of St. Christopher = St. Kitts, figure 5, number 35.
Tichosina dubia, n. sp.

155   Latitude 16° 41′ 54″ N, longitude 62° 13′ 24″ W, at 88 fathoms (= 161 meters), west off Montserrat, figure 5, number 36.
Tichosina sp. 8

157   Latitude 16° 41′ 54″ N, longitude 62° 13′ 24″ W at 120 fathoms (= 220 meters), west off Montserrat, figure 5, number 37.
Tichosina? bartletti (Dall)

167   Latitude 16° 09′ 40″ N, longitude 61° 29′ 25″ at 175 fathoms (= 320 meters), south side of Guadeloupe, figure 5, number 38.
Tichosina cubensis (Pourtalès)
T. sp. 8

193   Latitude 14° 43′ 48″ N, longitude 61° 11′ 25″ W, at 169 fathoms (= 309 meters), on the west side of Martinique, figure 3, number 41; figure 5, number 39.
Tichosina martinicensis (Dall)

232   Latitude 13° 06′ 45″ N. longitude 61° 06′ 55″ W, at 87 fathoms (= 159 meters) off southeast side of St. Vincent, figure 5, number 40. ber 40.
Tichosina dubia, n. sp.

253   Latitude 11° 25′ 00″ N, longitude 62° 04′ 15″ W at 96 fathoms (= 176 meters), south-southwest of Grenada, figure 5, number 41.
Tichosina sp. 3

19

254   Latitude 11° 27' 00" N, longitude 62° 11' 00" W at 164 fathoms (= 300 meters), south-southwest of Grenada, figure 5, number 42.

Tichosina dubia, n. sp.

CALIFORNIA INSTITUTE OF TECHNOLOGY STATION

1494   Off Gibbs Hill Lighthouse, Bermuda Platform, at 400 fathoms (= 732 meters). Not on map.

Notozyga lowenstami, n. sp.

COMBAT STATIONS

449   Latitude 24° 05' N, longitude 79° 46' W at 350 fathoms (= 641 meters), north of La Isabela, Cuba, figure 4, number 69.

Tichosina erecta, n. sp.

450   Latitude 23° 59' N, longitude 79° 43' W at 350 fathoms (= 641 meters), north of La Isabela, Cuba, figure 4, number 70.

Terebratulina cailleti Crosse
Tichosina erecta, n. sp.
T. elongata, n. sp.
T. sp. 5

EL TORITO STATION

NGS-
MRF-33   Bahama Islands: off Small Hope Bay, Goat Cay. Bearings: roof of Small Hope Bay Lodge, south end of Goat Cay; and lighthouse on tip of third cay south of Goat Cay. Depth, 55 to 150 feet. Habitat: sunken barge covered with coral and sea whips; bottom heavily covered with coral; reef drop-off covered with sponges and sea-whips. Cruise 4, April 20, 1970, coll. J. B. Starck, W. A. Starck, Peter Hopper. Not on map.

Thecidellina barretti (Davidson)
Argyrotheca sp., not described
A. schrammi Crosse and Fischer
Platidia sp., not described

EOLIS STATIONS

1   Sand Key, Florida, 125 fathoms (= 229 meters), 3 miles southsoutheast of Channel buoy. Not on map.

Tichosina solida, n. sp.
T. abrupta, n. sp.

31   Off Key West Florida at 90 fathoms (= 165 meters). Not on map.

Tichosina abrupta, n. sp.

340   Off Fowey Light, 209 fathoms, (= 382 meters), Florida Keys. Not on map.

Cryptopora rectimarginata Cooper

FISH HAWK STATIONS

6070   Mayaguez Harbor, E 5/8, S, 9 miles at 220 to 225 fathoms (= 403 to 412 meters), Puerto Rico, figure 5, number 44.

Tichosina subtriangulata, n. sp.

7283   Latitude 24° 17′ 05″ N, longitude 81° 53′ 30″ W at 127 fathoms (= 232 meters), in the Gulf Stream off Key West, Florida, figure 4, number 71.

Tichosina cubensis (Pourtalès)

GERDA STATIONS

G177   Latitude 27° 17′ N, longitude 79° 34′ W to 27° 20′ N, 79° 34′ W at 686 meters, northwest of west end of Grand Bahama, figure 4, number 14.

Tichosina aff. *T. bahamiensis,* n. sp.

G190   Latitude 25° 57′ N, longitude 78° 07′ W at 733-897 meters, south of Grand Bahama, figure 4, number 15.

Terebratulina cailleti Crosse
Platidia davidsoni (Deslongchamps)

G241   Latitude 25° 26.7′ N, longitude 79° 18′ W to 25° 33′ N, 79° 21′ W at 494-502 meters, southeast of Miami, Straits of Florida, figure 4, number 16.

Tichosina cubensis (Pourtalès)
Dallina floridana (Pourtalès)

G242   Latitude 25° 36′ N, longitude 79° 21′ W to 25° 38′ N, 79° 22′ W at 458-531 meters, southeast of Miami, Straits of Florida, figure, 4 number 17.

Tichosina cubensis (Pourtalès)
Dallina floridana (Pourtalès)

G246   Latitude 26° 57′ N, longitude 79° 12.5′ W to 27° 00′ N, 79° 12.5′ W at 512 meters, northwest of Grand Bahama Island, Little Bahama Bank, figure 4, number 18.

Dallina floridana (Pourtalès)

G252   Latitude 27° 29.5′ N, longitude 78° 37.5′ W at 485-496 meters, north of Grand Bahama Island, Little Bahama Bank, figure 4, number 19.

Tichosina rotundovata, n. sp.

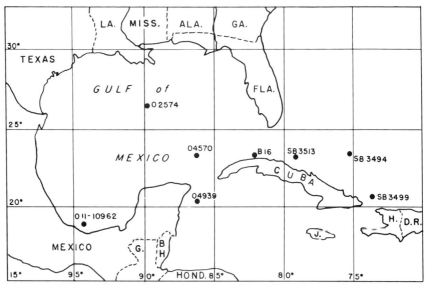

FIGURE 2. Map of the Caribbean Sea showing location of some BLAKE, OREGON and SILVER BAY stations.

G289   Latitude 24° 11′ N, longitude 81° 36′ W to 25° 15′ N, 81° 20′ W, at 604-594 meters in Straits of Florida, figure 4, number 20.
        Cryptopora rectimarginata Cooper

G295   Latitude 25° 13.5′ N, longitude 79° 27′ W to 25° 36′ N, 79° 23′ W, at 844-833 meters, Straits of Florida, southeast of Miami, figure 4, number 21.
        Tichosina sp. (immature)

G304   Latitude 25° 26′ N, longitude 79° 23′ W to 25° 30′ N, 79° 24′ W at 796 meters, southeast of Miami, Straits of Florida, figure 4, number 22.
        Platidia davidsoni (Deslongchamps)
        Terebratulina cailleti Crosse

G482   Latitude 24° 29′ N, longitude 80° 54′ W to 24° 32′ N, 80° 48′ W at 201-210 meters, Straits of Florida, east of Key West, figure 4, number 23.
        Terebratulina cailleti Crosse
        Tichosina rotundovata, n. sp.
        Dallina floridana (Pourtalès)

G503   Latitude 26° 31′ N, longitude 78° 51′ W to 26° 28′ N, 78° 45′ W at 366 meters, southwest side Grand Bahama Island, figure 4, number 24.
        Tichosina rotundovata, n. sp.

22

G533 Latitude 26° 27′ N, longitude 78° 43′ W to 26° 28′ N, 78° 45′ W at 384-403 meters, southwest side Grand Bahama Island, figure 4, number 25.

Argyrotheca barrettiana (Davidson)

G579 Latitude 24° 25′ N, longitude 81° 30′ W at 170-179 meters, Straits of Florida, southeast of Key West, Florida, figure 4, number 26.

Tichosina cubensis (Pourtalès)
Dallina floridana (Pourtalès)

G636 Latitude 26° 04′ N, longitude 79° 13′ W to 26° 04′ N, 79° 13′ W at 128-146 meters, east-northeast of Miami, Florida, figure 4, number 27.

Argyrotheca barrettiana (Davidson)

G646 Latitude 25° 49′ N, longitude 79° 21′ W to 25° 53′ N, 79° 22′ W at 439-531 meters, east-northeast of Miami, Florida, Straits of Florida, figure 4, number 28.

Erymnia muralifera, n. sp.

G678 Latitude 25° 57′ N, longitude 78° 13′ W to 25° 56′ N, 78° 09′ W at 540-576 meters, south of Grand Bahama Island, figure 4, number 29.

Cryptopora rectimarginata Cooper

G688 Latitude 26° 35′ N, longitude 78° 15′ W to 26° 35′ N, 78° 16′ W at 512-472 meters, south side Grand Bahama Island, figure 4, number 30.

Tichosina bahamiensis, n. sp.
T. cubensis (Pourtalès)

G691 Latitude 26° 35′ N, longitude 78° 24′ W to 26° 35′ N, 78° 25′ W at 333-375 meters, south side Grand Bahama Island, figure 4, number 31.

Cryptopora rectimarginata Cooper
Erymnia muralifera, n. sp.

G692 Latitude 26° 35′ N, longitude 78° 25′ W to 26° 34′ N, 78° 26′ W at 329-421 meters, south side of Grand Bahama Island, figure 4, number 32.

Terebratulina cailleti Crosse

G694 Latitude 26° 28′ N, longitude 78° 40′ W to 26° 27′ N, 78° 43′ W at 622-695 meters, south side Grand Bahama Island, figure 4, number 33.

Tichosina erecta, n. sp.

G695 Latitude 26° 28′ N, longitude 78° 37′ W to 26° 28′ N, 78° 43′ W at 555-575 meters, south side Grand Bahama Island, figure 4, number 34.

Erymnia muralifera, n. sp.

G701   Latitude 26° 29' N, longitude 78° 40' W at 311-275 meters, south side of Grand Bahama, figure 4, number 35.

Tichosina sp. indet.

G704   Same as above but at 366 to 275 meters, figure 4, number 35.

Erymia muralifera, n. sp.
Tichosina sp. 9

G706   Latitude 26° 27' N, longitude 78° 43' W to 26° 28' N, 78° 40' W at 522-489 meters, south side Grand Bahama Island, figure 4, number 36.

Erymnia muralifera, n. sp.

G707   Latitude 26° 27' N, longitude 78° 40' W, 26° 27' N, 78° 42' W at 514-586 meters at southwest corner Grand Bahama Island, figure 4, number 5.

Tichosina obesa n. sp.

G708   Latitude 26° 27' N, longitude 78° 46' W to 26° 27' N, 78° 47' W at 650 meters, south of Grand Bahama Island, figure 4, number 65.

Tichosina truncata, n. sp.

G713   Latitude 25° 59' N, longitude 79° 15' W to 25° 59' N, 79° 15' W at 190-201 meters, Straits of Florida, northeast of Miami, Florida, figure 4, number 37.

Terebratulina latifrons Dall

G816   Latitude 24° 04' N, longitude 79° 42' W at 558-540 meters, west of Andros Island, figure 4, number 38.

Terebratulina cailleti Crosse

G835   Latitude 24° 22' N, longitude 81° 11' W at 198-187 meters, Straits of Florida, southeast of Key West, figure 4, number 39.

Dallina floridana (Pourtalès)

G838   Latitude 24° 25' N, longitude 80° 58' W at 210-229 meters, Straits of Florida, southeast of Key West, figure 4, number 40.

Dallina floridana (Pourtalès)

G839   Latitude 24° 23' N, longitude 80° 52' W at 236-229 meters, southeast of Key West, Straits of Florida, figure 4, number 41.

Dallina floridana (Pourtalès)

G863   Latitude 24° 19' N, longitude 81° 07' W at 234 meters, Straits of Florida, southeast of Key West, figure 4, number 42.

24

Tichosina sp. indet. (dorsal fragment)
Dallina floridana (Pourtalès)

G864 Latitude 24° 18′ N, longitude 81° 07′ W at 223 meters. Straits of Florida, southeast of Key West, figure 4, number 43.

Dallina floridana (Pourtalès)

G898 Latitude 21° 04′ N, longitude 86° 19′ W at 338-366 meters, off Isla Mujeres, southwest side Yucatan Channel, Mexico, figure 4, number 44.

Tichosina truncata, n. sp.

G938 Latitude 26° 19′ N, longitude 79° 00′ W at 503-494 meters, northeast of Miami, Florida, figure 4, number 45.

Tichosina erecta, n. sp.

G942 Latitude 21° 10′ N, longitude 86° 21′ W at 494-490 meters, southwest side Yucatan Channel, northeast of Porto de Morelos, Mexico, figure 4, number 46.

Tichosina rotundovata, n. sp.
Dallina floridana (Pourtalès)

G944 Latitude 21° 05′ N, longitude 86° 21′ W (no depth given), southwest side Yucatan Channel, northeast of Porto de Morelos, Mexico, figure 4, number 47.

Tichosina sp. indet.

G947 Latitude 21° 13′ N, longitude 86° 25′ W at 284-247 meters, southwest side Yucatan Channel, northeast of Porto de Morelos, Mexico, figure 4, number 48.

Terebratulina latifrons Dall
Tichosina truncata, n. sp.

G972 Latitude 24° 24′ N, longitude 80° 52′ W at 231-221 meters, Straits of Florida, southeast of Key West, Florida, figure 4, number 49.

Dallina floridana (Pourtalès)

G973 Latitude 24° 22′ N, longitude 80° 55′ W at 329-275 meters, Straits of Florida, southeast of Key West, Florida, figure 4, number 50.

Tichosina cubensis (Pourtalès)
Dallina floridana (Pourtalès)

G974 Latitude 24° 22′ N, longitude 80° 57′ W at 253-251 meters, Straits of Florida, southeast of Key West, Florida, figure 4, number 51.

Tichosina cubensis (Pourtalès)
Dallina floridana (Pourtalès)

G982 Latitude 24° 05′ N, longitude 80° 20′ W at 130-90 fathoms (= 239-165 meters), southeast of Key West, Florida, figure 4, number 52.

Terebratulina latifrons Dall

25

G983   Latitude 24° 05' N, longitude 80° 20' W at 118 fathoms (= 217 meters) southeast of Key West, Florida, figure 4, number 53.

Terebratulina latifrons Dall
Tichosina sp. (fragment of dorsal valve)
Argyrotheca barrettiana (Davidson)

G984   Latitude 24° 05' N, longitude 80° 20' W at 85-126 fathoms (= 164-230 meters) southeast of Key West, Florida, figure 4, number 54.

Argyrotheca barrettiana (Davidson)

G985   Latitude 24° 06' N, longitude 80° 12' W at 201-37 meters, Straits of Florida, southeast of Key West, Florida, figure 4, number 55.

Terebratulina latifrons Dall
Tichosina sp. indet, (Fragment)
Argyrotheca barrettiana (Davidson)

G986   Latitude 24° 05' N, longitude 80° 19' W at 242-137 meters, Straits of Florida, southeast of Key West, Florida, figure 4, number 56.

Terebratulina latifrons Dall
Argyrotheca barrettiana (Davidson)

G1012   Latitude 23° 35' N, longitude 79° 33' W to 23° 37' N, 79° 32' W at 509-531 meters, northeast of La Isabela, Cuba, figure 4, number 57.

Tichosina ? bartletti (Dall)

G1029   Latitude 24° 17' N, longitude 81° 11' W at 291-302 meters, Straits of Florida, southeast of Key West, Florida, figure 4, number 58.

Tichosina solida, n. sp.
Dallina floridana (Pourtalès)

G1036   Latitude 24° 22.5' N, longitude 80° 53' W at 229-238 meters, Straits of Florida, southeast of Key West, Florida, figure 4, number 59.

Tichosina cubensis (Pourtalès)
Dallina floridana (Pourtalès)

G1082   Latitude 24° 24.5' N, longitude 82° 02.5' W at 115 meters southwest of Key West, Florida, figure 4, number 60.

Tichosina sp. (broken ventral valve).

G1102   Latitude 24° 15.5' N, longitude 81° 34' W at 247-283 meters, Straits of Florida, south of Key West, Florida, figure 4, number 61.

Tichosina cubensis (Pourtalès)
Dallina floridana (Pourtalès)

G1125   Latitude 26° 45′ N, longitude 79° 05′ W at 494-531 meters, off west end Grand Bahama Island, Straits of Florida, figure 4, number 62.

Tichosina erecta, n. sp.
Dallina floridana (Pourtalès)

G1312   Latitude 26° 40.5′ N, longitude 79° 04′ W to 26° 38.4′ N, 79° 02.5′ W at 512-503 meters off Miami, Florida, figure 4, number 63.

Tichosina rotundovata, n. sp.
Dallina floridana (Pourtalès)

G1314   Latitude 26° 52.4′ N, longitude 79° 11.8′ W, at 534-531 meters, north-northwest of Grand Bahama Island, figure 4, number 64.

Tichosina rotundovata, n. sp.
Dallina floridana (Pourtalès)

GERDA-
VII-1958   3 miles east of Golding Key, Tongue of the Ocean, Bahama Islands, at 500 fathoms (= 915 meters). Not on map.

Argyrotheca sp. 3

HYDRO STATION

5809   Latitude 33° 38.5′ N, longitude 73° 50.5′ W at 2500 to 2590 meters, off Cape Fear, South Carolina, figure 1.

Abyssothyris atlantica, n. sp.

JOHNSON-SMITHSONIAN DEEP SEA EXPEDITION STATIONS

43   Latitude 18° 02′ N, longitude 67° 51′ 14″ W to 18° 03′ 45″ N, 67° 48′ 10″ W at 240-300 fathoms (= 439-549 meters) near Mona Island, between Puerto Rico and the Dominican Republic. Not on map.

Stenosarina parva n. sp.

52   Latitude 19° 10′ 25″ N, longitude 69° 20′ 55″ to 10° 10′ 25″ W, 69° 21′ 25″ W at 14-22 fathoms (= 25 to 40 meters), east off Cape Samaná, Dominican Republic, figure 3, number 18; figure 4, number 83.

Crania pourtalesi Dall
Argyrotheca johnsoni Cooper
Lacazella caribbeanensis n. sp.

102   Latitude 18⁰ 50′ 30″ N, longitude 64⁰ 33′ 00″ W to 18⁰ 51′ 00″ N, 64⁰ 33′ 00″ W at 90-500 fathoms (= 169-915 meters), Virgin Islands, figure 3, number 19.

Tichosina ? bartletti (Dall)
T. bartschi (Cooper)

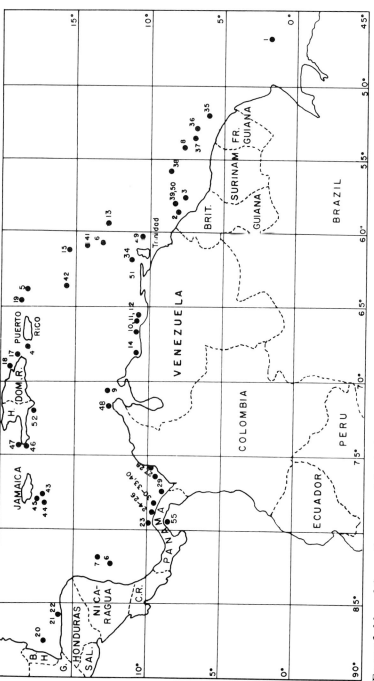

FIGURE 3. Map of the southern part of the Caribbean Sea showing stations of BLAKE, Johnson-Smithsonian Expedition, OREGON, PILLSBURY and SILVER BAY.

Figure 3. (Continued) Dredging Stations.

1. OREGON 2081	11. OREGON 4461	21. P629
2. OREGON 2232	12. OREGON 4466	22. P630
3. OREGON 2248	13. OREGON 5018	23. P325
4. OREGON 2644	14. OREGON 5624	24. P330
5. OREGON 2646	15. OREGON 5927	25. P331
6. OREGON 3608	16. OREGON 5955	26. P338
7. OREGON 6423	17. SILVER BAY 5181	27. P365
8. OREGON 4297	18. Johnson-	28. P389
9. OREGON 4398	Smithsonian 52	29. P409
10. OREGON 4459	19. Johnson-	30. P415
	Smithsonian 102	
	20. P610	
31. P419	41. BLAKE 193	51. P707
32. P420	42. OREGON 4994	52. P1266
33. P422	43. P1262	53. P566, P572
34. P479	44. P1256	in Balboa Harbor
35. P650	45. P1225	
36. P657	46. P1186	
37. P658	47. P1181	
38. P680	48. P769	
39. P694	49. P838	
40. P421	50. OREGON II 10513	

OREGON STATIONS

2081   Latitude 01° 52′ N, longitude 46° 54′ W at 175 fathoms (= 330 meters), east off Cape Raso, Brazil, figure 3, number 1.
Tichosina sp. 7

2232   Latitude 08° 31′ N, longitude 58° 37′ W at 48 fathoms (= 88 meters), off Waini Point, British Guiana, figure 3, number 2.
Tichosina obesa, n. sp.

2248   Latitude 07° 45′ N, longitude 57° 34″ W at 35-30 fathoms (= 64-55 meters), northeast of Georgetown, British Guiana, figure 3, number 3.
Terebratulina latifrons Dall

2574   Latitude 26° 34′ N, longitude 89° 53′ W at 1450 fathoms (= 2653 meters), north side of Mexican Basin, Gulf of Mexico, figure 2; figure 4, number 80.

Chlidonophora incerta (Davidson)

29

2644   Latitude 18° 14' N, longitude 67° 42' W at 260 fathoms (= 476 meters) west side of Puerto Rico, figure 3, number 4.
Dallina floridana (Pourtalès)

2646   Latitude 18° 14' N, longitude 64° 20' W at 210 fathoms (= 384 meters), Virgin Islands, figure 3, number 5.
Dallina floridana (Pourtalès)

3608   Latitude 12° 28' N, longitude 82° 28' W at 110 fathoms (= 201 meters) east of Nicaragua, figure 3, number 6.
Tichosina bullisi, n. sp.

4297   Latitude 07° 46' N, longitude 54° 17' W at 350 fathoms (= 641 meters) north of Paramaribo, Surinam, figure 3, number 8.
Tichosina obesa, n. sp.

4398   Latitude 12° 46' N, longitude 70° 41' W at 110 fathoms (= 201 meters), east of Punta Gallinas, Colombia, figure 3, number 9.
Terebratulina cailleti Crosse

4459   Latitude 10° 50' N, longitude 66° 58' W at 53 fathoms (= 97 meters) north of Caracas, Venezuela, figure 3, number 10.
Terebratulina cailleti Crosse
Tichosina obesa, n. sp.

4461   Latitude 10° 50' N, longitude 66° 55' W, at 53 fathoms (= 97 meters) northeast of Caracas, Venezuela, figure 3, number 11.
Tichosina obesa, n. sp.

4466   Latitude 10° 44' N, longitude 66° 09' W at 40 fathoms (= 73 meters) northeast of Caracas, Venezuela, figure 3, number 12.
Tichosina obesa, n. sp.

4928   Latitude 14° 05' N, longitude 81° 21' W at 100 fathoms (= 183 meters) off Nicaragua. Not on map.
Tichosina sp. 4

4939   Latitude 20° 25' N, longitude 86° 13' at 150 fathoms (355 = meters) off Isla de Cozumel, Mexico, figure 2; figure 4, number 82.
Stenosarina cf. *S. angustata* n. sp.

4994   Latitude 15° 30' N, longitude 63° 38' W at 220-205 fathoms (= 366 to 375 meters), southeast of the Island of Aves, figure 3, number 42.
Small, indeterminate *Tichosina* ventral valve
Dallina floridana (Pourtalès)

5015   Latitude 13° 02' N, longitude 59° 34' W at 110-135 fathoms (= 201-247 meters), south side of Barbados. Not on map.
Tichosina sp. 6

5018   Latitude 13° 00′ N, longitude 59° 33′ W at 175 fathoms (= 330 meters), south side of Barbados, figure 3, number 13; figure 5, number 43.

Terebratulina cailleti Crosse

5624   Latitude 10° 52′ N, longitude 68° 08′ W at 56 fathoms (= 102 meters), north of Puerto Cabello, Venezuela, figure 3, number 14; figure 5, number 7.

Tichosina plicata, n. sp.

5927   Latitude 15° 36′ N, longitude 61° 13′ W at 332 fathoms (= 608 meters) on the northeast side of the Island of Dominica, figure 3, number 15.

Tichosina cubensis (Pourtalès)
Stenosarina angustata n. sp.
S. nitens n. sp.

5955   Latitude 13° 14′ N, longitude 60° 53′ W at 90 fathoms (= 165 meters) east side of St. Vincent, figure 3, number 16.

Terebratulina latifrons Dall

6423   Latitude 13° 28′ N, longitude 82° 01′ W at 158 fathoms (= 289 meters) off Nicaragua, figure 3, no. 7.

Tichosina bullisi, n. sp.

6715   Latitude 18° 36′ N, longitude 63° 27′ W at 110-130 fathoms (= 201-238 meters), Virgin Islands, figure 5, number 45.

Tichosina sp. 1, 2.

OREGON II STATIONS

10513   Latitude 08° 26′ N, longitude 58° 11′ W at 100 fathoms (= 183 meters), north of Georgetown, British Guiana, figure 3, number 50.

Tichosina dubia, n. sp.

PILLSBURY STATIONS

P120   Latitude 31° 48′ N, longitude 76° 38′ W to 31° 49′ N, 76° 26′ W at 2194-2373 meters east of Savannah, Georgia, figure 1.

Pelagodiscus atlanticus (W. King)

P325   Latitude 9° 52′ N, longitude 79° 35.5′ W to 9° 52′ N, 79° 37′ W at 1794-1656 meters, northeast of Punta Manzanillo, Panama, figure 3, number 23.

Chlidonophora incerta (Davidson)

P330   Latitude 9° 37.5′ N, longitude 78° 54′ W to 9° 37.0′ N, 78° 52.5′ W at 127-63 meters, northeast of Punta de San Blas, Panama, figure 3, number 24.

Argyrotheca barrettiana (Davidson)

31

P331 Latitude 9° 31′ N, longitude 78° 56′ W to 9° 31.5′ N, 78° 55.2′ W at 26-46 meters, northeast of Punta de San Blas, Panama, figure 3, number 25.

Argyrotheca rubrocostata n. sp.

P338 Latitude 9° 57.5′ N, longitude 78° 31′ W to 9° 58.3′ N, 78° 30.5′ W at 1836-1822 meters, northeast of Punta de San Blas, Panama, figure 3, number 26.

Pelagodiscus atlanticus (King)

P365 Latitude 9° 31.3′ N, longitude 76° 15.4′ W to 9° 32.5′ N, 76° 17′ W at 56-58 meters, northwest of Punta Piedras, Colombia, figure 3, number 27.

Platidia anomioides (Scacchi and Philippi)

P389 Latitude 9° 53.8′ N, longitude 75° 50.9′ W to 9° 54.2′ N, 75° 52.4′ W at 70-51 meters, northwest of Punta San Bernardo, Colombia, figure 3, number 28.

Argyrotheca barrettiana (Davidson)

P409 Latitude 8° 51.2′ N, longitude 77° 28.1′ W to 8° 51.2′ N, 77° 28.1′ W at 55-48 meters, north-northwest of Cape Tiburon, northwest of the Gulf of Darien, Colombia, figure 3, number 29.

Argyrotheca barrettiana (Davidson)

P415 Latitude 9° 22.4′ N, longitude 78° 08.4′ W to 9° 22.4′ N, 78° 08.4′ W at 60 meters, east-northeast of Punta de San Blas, Panama, figure 3, number 30.

Argyrotheca barrettiana (Davidson)?

P419 Latitude 9° 28.3′ N, longitude 78° 20.7′ W to 9° 29.2′ N, 78° 21.7′ W at 51-55 meters, same as above, figure 3, number 31.

Argyrotheca rubrocostata, n. sp.

P420 Latitude 9° 30.5′ N, longitude 78° 25.6′ W to 9° 30.7′ N, 78° 26.0′ W at 28 meters, same as above, figure 3, number 32.

Argyrotheca rubrocostata, n. sp.

P421 Latitude 9° 32.1′ N, longitude 78° 33.5′ W, 9° 32.5′ N, 78° 34.3′ W at 53-59 meters, east of Punta de San Blas, Panama, figure 3, number 40.

Argyrotheca barrettiana (Davidson)

P422 Latitude 9° 33.8′ longitude 78° 36.2′ W to 9° 34.7′ N, 78° 36.7′ W at 73-70 meters, east-northeast of Punta de San Blas, Panama, figure 3, number 33.

Argyrotheca barrettiana (Davidson)

P479 Latitude 11° 19.7′ N, longitude 62° 03.1′ W to 11° 19.7′ N, 62° 01.2′

W at 122-134 meters, north-northwest of Port of Spain, Trinidad, figure 3, number 34.

Tichosina obesa, n. sp.

P566   On mud flats under east side of Bridge of the Americas, Balboa Harbor, Panama at low tide, figure 3, number 55.

Glottidia audebarti (Broderip)

P572   On mud flats one-half mile south of the Bridge of the Americas west side Balboa Harbor, Panama, figure 3, number 55.

Glottidia audebarti (Broderip)

P584   Latitude 21° 02′ N, longitude 86° 24′ W at 353-348 meters southeast of Mujeres Island, on the southwest side of the Yucatan Channel, Mexico, figure 4, number 1.

Terebratulina cailleti (Crosse)
Tichosina truncata, n. sp.
T. obesa, n. sp.
T. expansa, n. sp.

P587   Latitude 21° 17′ N, longitude 86° 13′ W at 412-458 meters, east of Mujeres Island, southwest side of Yucatan Channel, Mexico, figure 4, number 2.

Tichosina truncata, n. sp.
Dallina floridana (Pourtalès)
Platidia anomioides (Scacchi and Philippi)

P594   Latitude 21° 00.5′ N, longitude 86° 23′ W at 308-329 meters, southwest side Yucatan Channel, northeast of Porto de Morelos, Mexico, figure 4, number 3.

Tichosina truncata, n. sp.
Terebratulina cailleti Crosse
Platidia anomioides (Scacchi and Philippi)

P600   Latitude 20° 29′ N, longitude 87° 06′ W at 439-462 meters, south side of Isla de Cozumel, Mexico, figure 4, number 4.

Terebratulina cailleti Crosse (very large ?)

P610   Latitude 17° 02.0′ N, longitude 87° 38.4″ W to 17° 03.0′ N, 87° 38.5′ W at 296-329 meters, southeast of Turneffe Island, British Honduras, figure 3, number 20.

Indeterminate terebratulid fragments.

P629   Latitude 15° 58.2′ N, longitude 86° 09.0′ W to 15° 58.0′ N, 86° 10.0′ W at 40 meters, off Trujillo, Honduras, figure 3, number 21.

Argyrotheca rubrocostata, n. sp.

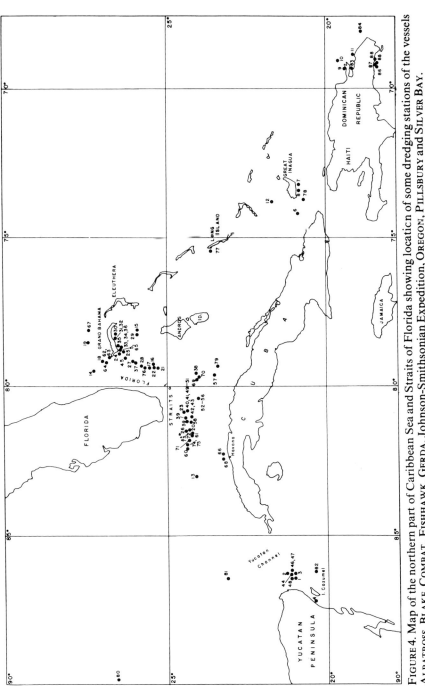

FIGURE 4. Map of the northern part of Caribbean Sea and Straits of Florida showing location of some dredging stations of the vessels ALBATROSS, BLAKE, COMBAT, FISHHAWK, GERDA, Johnson-Smithsonian Expedition, OREGON, PILLSBURY and SILVER BAY.

Figure 4. (Continued) Dredging Stations.

1. P584	11. P1158	21. G295
2. P587	12. P1434	22. G304
3. P594	13. G114	23. G482
4. P600	14. G177	24. G503
5. G707, ca. 65	15. G190	25. G533
6. P1138	16. G241	26. G579
7. P1141	17. G242	27. G636
8. P1142	18. G246	28. G646
9. P1153	19. G252	29. G678
10. P1157	20. G289	30. G688

31. G691	41. G839	51. G974
32. G692	42. G863	52. G982
33. G694	43. G864	53. G983
34. G695	44. G898	54. G984
35. G701, G704	45. G938	55. G985
36. G706	46. G942	56. G986
37. G713	47. G944	57. G1012
38. G816	48. G947	58. G1029
39. G835	49. G972	59. G1036
40. G838	50. G973	60. G1082

61. G1102	71. FISHHAWK 7283	81. OREGON 4570
62. G1125	72. SILVER BAY 2416	82. OREGON 4939
63. G1312	73. SILVER BAY 2418	83. Johnson-Smithsonian
64. G1314	74. SILVER BAY 2426	52
65. G708 = ca. 5	75. SILVER BAY 2427	84. SILVER BAY 5181
66. ALB 2343	76. SILVER BAY 2488	85. P1387
67. ALB 2655	77. SILVER BAY 3494	86. P1393
68. BLAKE 16	78. SILVER BAY 3499	87. P1303
69. COMBAT 449	79. SILVER BAY 3513	88. P1386
70. COMBAT 450	80. OREGON 2574	

P630   Latitude 15° 59.2′ N, longitude 86° 02′ W to 15° 59.2′ N, 86° 04.5′ W at 34-37 meters, off Trujillo, Honduras, figure 3, number 22.
Argyrotheca barrettiana (Davidson)
A. rubrocostata, n. sp.
A. sp. 2
A. sp. indet.
Thecidellina aff. *T. barretti* (Davidson)

P650   Latitude 06° 07′ N, longitude 52° 19′ W at 84-92 meters, north of Cayenne, French Guiana, figure 3, number 35.
Tichosina obesa, n. sp. ?

P657   Latitude 06° 58′ N, longitude 53° 10′ W, to 07° 01′ N, 53° 15′ W, at 128-132 meters, north of Mana, French Guiana, figure 3, number 36.

Tichosina ? aff. *T.* ? *bartletti* (Dall)

P658   Latitude 07° 10′ N, longitude 53° 36′ W at 135-123 meters, north of Mana, French Guiana, figure 3, number 37.

Terebratulina cailleti Crosse
Tichosina dubia, n. sp. ?

P680   Latitude 08° 42′ N, longitude 55° 48′ W to 09° 17′ N, 55° 45′ W at 3115-3239 meters, northeast of Georgetown, British Guiana, figure 3, number 38.

Chlidonophora incerta (Davidson)

P694   Latitude 08° 28′ N, longitude 58° 12′ W at 79-82 meters, north of Georgetown, British Guiana, figure 3, number 39.

Tichosina obesa, n. sp.

P707   Latitude 11° 21′ N, longitude 62° 21′ W to 11° 23.5′ N, 62° 23.0′ W at 79 meters, north of Punta Peñas, Venezuela, figure 3, number 51.

Terebratulina cailleti Crosse
Tichosina obesa, n. sp.

P708   Latitude 11° 24.7′ N, longitude 62° 40.5′ W to 11° 26.6′ N, 62° 40.5′ W at 70-73 meters, west-northwest of Port of Spain, Trinidad, figure 5, number 1.

Terebratulina cailleti Crosse
Tichosina obesa, n. sp.

P734   Latitude 11° 01.8′ N, longitude 65° 34.2′ W to 11° 01.0′ N, 65° 36.3′ W at 68-60 meters, west side of Isla la Tortuga, Cariaco Trench, northwest of Guanta, Venezuela, figure 5, number 2.

Terebratulina cailleti Crosse
Tichosina obesa, n. sp.

P736   Latitude 10° 57′ N, longitude 65° 52′ W to 11° 03′ N, 65° 59′ W at 70-156 meters, northeast of Caracas, Venezuela, figure 5 number 3.

Tichosina obesa, n. sp.
Platidia anomioides (Scacchi and Philippi)

P737   Latitude 10° 44′ N, longitude 66° 07′ W to 10° 45′ N, 66° 08′ W at 60-73 meters, northeast of Caracas, Venezuela, figure 5, number 4.

Terebratulina cailleti Crosse
Tichosina obesa, n. sp.

P739	Latitude 10° 54.7' N, longitude 66° 17.8' W to 10° 57.6' N, 66° 18.0' W at 254-280 meters, northeast of Caracas, Venezuela, figure 5, number 5.

Terebratulina cailleti Crosse
Megerlia echinata (Fisher and Oehlert)
Tichosina obesa, n. sp.

P741	Latitude 11° 47.8' N, longitude 66° 06.8' W to 11° 52.4' N, 66° 14.0' W at 1052-1067 meters, north of Isla Orchila, Venezuela, figure 5, number 6.

Unidentifiable terebratulid fragment = *Tichosina* ?

P769	Latitude 12° 31.0' N, longitude 71° 41.0' W to 12° 32.2' N, 71° 39.4' W at 133-146 meters, off Punta Gallinas, Colombia, figure 3, number 48.

Terebratulina latifrons Dall

P838	Latitude 10° 32' N, longitude 60° 23' W at 93-115 meters off the east side of Trinidad, figure 3, no. 49; figure 5, number 8.

Tichosina plicata, n. sp. ?
Terebratulina cailleti Crosse

P848	Latitude 11° 22' N, longitude 61° 26.4' W to 11° 23.8' N, 61° 25.8' W at 146 meters off the north side of Trinidad, figure 5, number 9.

Terebratulina cailleti Crosse
A. sp. indet.

P849	Latitude 11° 14.5' N, longitude 61° 46.2' W at 137-143 meters, off the north side of Trinidad, figure 5, number 10.

Tichosina sp. indet. (broken valves)
Terebratulina cailleti Crosse

P854	Latitude 12° 02' N, longitude 61° 35.7' W, at 73 meters on the south side of Grenada. figure 5, number 48.

Argyrotheca sp. aff. *A. rubrotincta* (Dall)?
A. sp. indet.

P855	Latitude 12° 07' N, longitude 61° 21.6' W at 27 to 29 meters east side of Grenada, figure 5, number 11.

Argyrotheca bermudana Dall

P856	Latitude 12° 17.5' N, longitude 61° 29' W at 35 meters northeast of Grenada, figure 5, number 12.

Argyrotheca bermudana Dall

P857	Latitude 12° 23.5' N, longitude 61° 21.6' W at 20-137 fathoms (= 37-251 meters) off the northeast side of Grenada, figure 5, number 13.

Argyrotheca crassa, n. sp.

37

P861   Latitude 12° 42′ N, longitude 61° 05.5′ W to 12° 42.5′ N, 61° 07.3′ W at 18-744 meters, northeast of Grenada, figure 5, number 14.

Tichosina sp. indet. (fragments)
Terebratulina cailleti Crosse
Platidia anomioides (Scacchi & Philippi)

P867   Latitude 13° 03′ N, 61° 06.8′ W to 13° 03.2′ N, 61° 06.7′ W at 37 meters, southeast of St. Vincent, figure 5, number 15.

Argyrotheca sp. immature, indet.

P871   Latitude 13° 14′ N, longitude 61° 30′ W to 13° 19.2′ N, 61° 28.6′ W at 2628-2681 meters, west of St. Vincent, figure 5, number 16.

Chlidonophora incerta (Davidson)

P874   Latitude 13° 11.2′ N, longitude 61° 05.3′ W to 13° 13.8′ N, 61° 05.1′ W at 156-201 meters, east side of St. Vincent, figure 5, number 17.

Tichosina indet. (broken valves)

P876   Latitude 13° 13.9′ N, longitude 61° 04.7′ W at 231-258 meters, east side of St. Vincent, figure 5, number 18.

Tichosina labiata, n. sp.
Terebratulina cailleti Crosse

P877   Latitude 13° 16.7′ N, longitude 61° 05.6′ W to 13° 18.6′ N, 61° 06.1′ W at 329-467 meters, northwest of St. Vincent, figure 5, number 19.

Platidia anomioides (Scacchi and Philippi)

P890   Latitude 14° 05.6′ N, longitude 60° 51.4′ W to 14° 07′ N, 60° 52.2′ W at 198-430 meters, off the northeast side of St. Lucia, figure 5, number 20.

Tichosina cubensis (Pourtalès)

P903   Latitude 13° 44′ N, longitude 61° 03.1′ W at 231-430 meters, off the southwest side of St. Lucia, figure 5, number 21.

Terebratulina cailleti Crosse

P905   Latitude 13° 46.3′ N, longitude 61° 05.4′ W at 384-963 meters, off the southwest side of St. Lucia, figure 5, number 22.

Tichosina cubensis (Pourtalès)
Platidia anomioides (Scacchi and Philippi)
Terebratulina cailleti Crosse

P907   Latitude 14° 26.8′ N, longitude 60° 58.3′ W at 115-214 meters, off the south side of Martinique, figure 5, number 23.

Tichosina ? bartletti (Dall)

P924   Latitude 16° 05′ N, longitude 61° 24′ W to 16° 06.2′ N, 61° 22.7′ W at 476-686 meters, off the southeast side of Guadeloupe, figure 5, number 24.

        Argyrotheca barrettiana (Davidson)

P929   Latitude 15° 29.5′ N, longitude 61° 11.5′ W, to 15° 30.5′ N, 61° 11.6′ W at 458-503 meters, off the east side of the Island of Dominica, figure 5, number 25.

        Tichosina cubensis (Pourtalès)
        Terebratulina cailleti Crosse
        Platidia anomioides (Scacchi and Philippi)

P930   Latitude 15° 29.7′ N, longitude 61° 12′ W to 15° 30.3′ N, 61° 12.4′ W at 210-399 meters, off the east side of the Island of Dominica, figure 5, number 26.

        Tichosina sp. aff. *T. cubensis* (Pourtalès)

P931   Latitude 15° 31.2′ N, longitude 61° 12.3′ W to 15° 32.0′ N, 61° 13.1′ W at 146-494 meters, off the east side of the Island of Dominica, figure 5, number 27.

        Tichosina cubensis (Pourtalès)

P943   Latitude 16° 25.9′ N, longitude 61° 36.7′ W to 16° 26.4′ N, 61° 36.4′ W at 275 meters, off the north side of Guadeloupe, figure 5, number 28.

        Fragments, probably *Tichosina* sp. aff. *T. cubensis* (Pourtales)
        Terebratulina cailleti Crosse

P954   Latitude 16° 55′ N, longitude 62° 43′ W to 16° 58.6′ N, 62° 46.5′ W at 686-1043 meters, southwest of Nevis, figure 5, number 29.

        Terebratulina cailleti Crosse

P976   Latitude 17° 30.6′ N, longitude 61° 23.7′ W to 17° 40′ N, 61° 18.5′ W at 3733-3971 meters, northeast of Antigua, figure 5, number 30.

        Pelagodiscus atlanticus (King)

P984   Latitude 18° 26.4′ N, longitude 63° 12.6′ W to 18° 28′ N, 63° 11.1′ W at 393-451 meters off the west side of Anguilla, figure 5, number 31.

        Tichosina cubensis (Pourtalès)?

P988   Latitude 18° 29.3′ N, longitude 63° 24.6′ W to 18° 31′ N, 63° 24.1′ W at 686-724 meters off the northwest side of Anguilla, figure 5, number 32.

        Dallina floridana (Pourtalès) (very young)

P991   Latitude 18° 47′ N, longitude 64° 46.8′ W at 205-380 meters, north of the Virgin Islands, figure 5, number 33.

        Tichosina cubensis (Pourtalès)

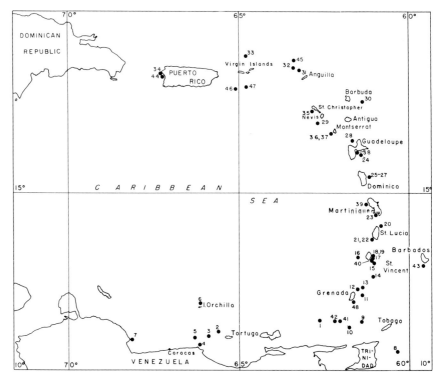

FIGURE 5. Map of the eastern Caribbean and Antilles showing location of dredging stations of Blake, Oregon and Pillsbury.

Erymnia muralifera, n. sp.
Argyrotheca barrettiana (Davidson)

P1138   Latitude 20° 51.7′ N, longitude 74° 22′ W to 20° 54.8′ N, 74° 18.6′ W at 2745-2751 meters, west of Great Inagua, figure 4, number 6.

Chlidonophora incerta (Davidson)
Abyssothyris? sp. 2

P1141   Latitude 20° 52′ N, longitude 73° 14′ W to 20° 50′ N, 73° 19′ W, at 402.6-457.5 meters, south side of Great Inagua, figure 4, number 7.

Tichosina sp. indet. (truncated anterior)

P1142   Latitude 20° 53′ N, longitude 73° 24′ W to 20° 53.6′ N, 73° 25.6′ W at 247.05-292.80 meters, southwest side of Great Inagua. (might = P1148, lat. 20° 00.4′ N, 71° 40.7′ W at 38 meters south of Caicos Island), figure 4, number 8.

Terebratulina latifrons Dall

40

Figure 5. (Continued) Dredging Stations.

1. P708	11. P855	21. P903
2. P734	12. P856	22. P905
3. P736	13. P857	23. P907
4. P737	14. P861	24. P924
5. P739	15. P867	25. P929
6. P741	16. P871	26. P930
7. OREGON 5624	17. P874	27. P931
8. P838	18. P876	28. P943
9. P848	19. P877	29. P954
10. P849	20. P890	30. P976

31. P984	41. BLAKE 253
32. P988	42. BLAKE 254
33. P991	43. OREGON 5018
34. Anasco Bay	44. Maguey Harbor
35. BLAKE 147	45. OREGON 6715
36. BLAKE 155	46. P1401
37. BLAKE 157	47. P1402
38. BLAKE 167	48. P854
39. BLAKE 193	
40. BLAKE 232	

P1153   Latitude 19° 22.2′ N, longitude 69° 19.3′ W, at 42.09-47.58 meters, just west of Cape Cabron, Dominican Republic, figure 4, number 9.
Argyrotheca johnsoni Cooper

P1157   Latitude 19° 06.3′ N, longitude 69° 01′ W, at 18.3-40.26 meters, northeast of Cape Samaná, Dominican Republic, figure 4, number 10.
Argyrotheca cf. *A. barrettiana* (Davidson)

P1158   Latitude 19° 03.1′ N, longitude 68° 47.2′ W, at 84.18-256.2 meters, near mouth of Samaná Bay, Dominican Republic, figure 4, number 11.
Terebratulina cailleti Crosse

P1181   Latitude 18° 51′ N, longitude 74° 30′ W to latitude 18° 46.8′ N, longitude 74° 35.9′ W at 2489-2548 meters, northwest of Jeremie, Haiti, figure 3, number 47.
Pelagodiscus atlanticus (W. King)

P1186   Latitude 18° 29.7′ N, longitude 74° 38.7′ W, at 183 meters, west of Jeremie, Haiti, figure 3, number 46.

Tichosina sp. (large ventral valves)

P1225   Latitude 17° 42.5′ N, longitude 77° 58.0′ W to latitude 17° 47′ N, longitude 77° 55.8′ W, at 457-558 meters, off the southwest side of Jamaica, figure 3, number 45.

Terebratulina cailleti Crosse

P1256   Latitude 17° 27′ N, longitude 78° 10′ W at 521-658 meters, off the south of east end of Jamaica, figure 3, number 44.

Terebratulina cailleti Crosse

P1262   Latitude 17° 21.4′ N, longitude 77° 34.8′ W at 805-1089 meters, off the south side of Jamaica, figure 3, number 43.

Dyscolia wyvillei (Davidson)

P1266   Latitude 17° 53′ N, longitude 71° 53′ W, at 1519-4173 meters, off the south side of Haiti, figure 3, number 52.

Chlidonophora incerta (Davidson)

P1303   Latitude 18° 21′ N, longitude 69° 14.3′ W at 174 meters, southeast of San Pedro de Macoris, Dominican Republic, figure 4, number 87.

Terebratulina cailleti Crosse

P1386   Latitude 18° 21.4′ N, longitude 69° 06′ W to 18° 22.8′ N, 69° 06.6′ W, at 148-92 meters, southeast of San Pedro de Macoris, Dominican Republic, figure 4, number 88.

Argyrotheca barrettiana (Davidson)

P1387   Latitude 18° 21.4′ N, longitude 69° 08.7′ W at 165-130 meters southeast of San Pedro de Macoris, Dominican Republic, figure 4, number 85.

Tichosina pillsburyae, n. sp.

P1393   Latitude 18° 21.7′ N, longitude 69° 18.4′ W at 150 meters, south of San Pedro de Macoris, Dominican Republic, figure 4, number 86.

Terebratulia cailleti Crosse
Tichosina pillsburyae, n. sp.

P1397   Latitude 17° 54.3′ N, longitude 65° 09.5′ W at 4352-4366 meters, northwest of St. Croix, southeast of east end of Puerto Rico, not on map.

Chlidonophora incerta (Davidson)

P1401   Latitude 17° 51′ N, longitude 65° 04.2′ W at 4226-4133 meters southeast of Puerto Rico, figure 5, number 46.

Chlidonophora incerta (Davidson)

P1402   Latitude 17° 55.3′ N, longitude 64° 47′ W at 3950-3936 meters, southeast of Puerto Rico, figure 5, number 47.

Chlidonophora incerta (Davidson)

P1421   Latitude 21° 36.1′ N, longitude 71° 01′ at 73-95 meters, northeast of Turks Island, Bahama Islands, not on map.

Argyrotheca johnsoni Cooper

P1434   Latitude 21° 40.7′ N, longitude 73° 50.8′ W at 18 meters, north-northwest of Great Inagua, figure 4, number 12.

Aryrotheca barrettiana (Davidson)
A. sp. 1

P1438   Latitude 22° 27.3′ N, longitude 73° 10.1′ W at 770-742 meters, northwest end of Mayaguana Island, Bahama Islands, not on map.

Tichosina sp. (detached ventral valves)

SILVER BAY STATIONS

2416   Latitude 24° 18′ N, longitude 81° 29′ W, at 125 fathoms (= 229 meters), Straits of Florida, figure 4, number 72.

Tichosina cubensis (Pourtalès)
Dallina floridana (Pourtalès)

2418   Latitude 24° 14′ N, longitude 81° 24′ W at 160 fathoms (= 293 meters), Straits of Florida, figure 4, number 73.

Tichosina cubensis (Pourtalès)
T. solida, n. sp.
Dallina floridana (Pourtalès)

2426   Latitude 24° 23′ N, longitude 81° 59′ W at 120 fathoms (= 220 meters), near Key West, Straits of Florida, figure 4, number 74.

Tichosina solida, n. sp.
Dallina floridana (Pourtalès)

2427   Latitude 24° 20′ N, longitude 82° 04′ W at 120 fathoms (= 220 meters), near Key West, Straits of Florida, figure 4, number 75.

Tichosina cubensis (Pourtalès)
Dallina floridana (Pourtalès)

2488   Latitude 25° 41′ N, longitude 79° 20′ W at 115 fathoms (= 210 meters) in Straits of Florida, southeast of Miami, Florida, figure 4, number 76.

Dallina floridana (Pourtalès)

3494   Latitude 23° 36′ N, longitude 75° 25′ W at 100-200 fathoms (= 183 to 366 meters), west side of north end of Long Island, Bahamas, figure 2; figure 4, number 77.

Erymnia angusta, n. sp.
Dallina floridana (Pourtalès)

3499	Latitude 20° 44′ N, longitude 73° 43′ W at 300 fathoms (= 549 meters) southwest of Great Inagua, figure 2; figure 4, number 78.

Tichosina expansa, n. sp.

3513	Latitude 23° 26′ N, longitude 79° 24′ W at 549-595 meters, north of La Isabela, Cuba, figure 2; figure 4, number 79.

Dallina floridana (Pourtalès)

5181	Latitude 18° 50′ N, longitude 68° 13.5′ W at 300 fathoms (= 549 meters), north of the east end of the Dominican Republic, figure 3, number 17; figure 4, number 84.

Dallina floridana (Pourtalès)
Stenosarina nitens, n. sp.

DREDGING STATIONS AND COLLECTIONS FROM THE GULF OF MEXICO
ALAMINOS STATIONS

4	Latitude 25° 08′ N, longitude 94° 48′ W at 1948 fathoms (= 3565 meters), on the Continental Rise* due south of Galveston, and southeast of Brownsville, Texas. Number 2 on map.

Chlidonophora incerta (Davidson)

4c	Latitude 23° 36′ N, longitude 93° 57′ W at 2105 fathoms (= 3852 meters) on the Abyssal Plain due south of Port Arthur, Texas, and northeast of Tampico, Mexico. Number 5 on map.

Chlidonophora incerta (Davidson)

5	Latitude 23° 44′ N, longitude 92° 36′ W at 2100 fathoms (= 3843 meters) on the Abyssal Plain northeast of Tampico, Mexico. Number 6 on map.

Chlidonophora incerta (Davidson)

9	Latitude 29° 27′ N, longitude 86° 57.1′ W at 400 plus fathoms (= 732 plus meters) on the West Florida Shelf due east of New Orleans, Louisiana. Number 15 on map.

Platidia anomioides (Scacchi and Philippi)

12	Latitude 23° 44.5′ N, longitude 92° 29.5′ W at 1900 fathoms (= 3477 meters) on the Abyssal Plain northeast of Tampico, Mexico. Number 7 on map.

Chlidonophora incerta (Davidson)

ALBATROSS STATIONS

2383	Latitude 28° 32′ N, longitude 88° 06′ W at 1181 fathoms (= 2161 meters) on the Mississippi Cone almost due south of Bay. Number 23 on map.

Chlidonophora incerta (Davidson)

*Terminology of bottom geography of the Gulf of Mexico taken from Harding and Nowlin, 1967. See references.

2388   Latitude 29° 24′ 30″ N, longitude 88° 10′ 00″ W at 35 fathoms (= 64 meters) on the West Florida Shelf south-southeast of Pascagoula, Mississippi. Number 14 on map.
        Dallina floridana (Pourtalès) — juvenile specimen

2400   Latitude 28° 41′ N, longitude 86° 07′ W at 67 fathoms (= 123 meters) on the Upper Continental Slope south-southwest of Panama City, Florida. Number 21 on map.
        Cryptopora rectimarginata Cooper

BLAKE STATIONS

45   Latitude 25° 33′ N, longitude 84° 21′ W at 101 fathoms (= 185 meters) on the West Florida Shelf, Bay of Florida. Number 18 on map.
        Argyrotheca barrettiana (Davidson)
        Tichosina floridensis, n. sp.

100   Off Morro Light, Cuba, at 400 fathoms (= 732 meters) not on map.
        Tichosina elongata, n. sp.

ROBERT COOPER COLLECTION

West-southwest of Dry Tortugas, Florida at 183 meters. DT on map.
        Terebratulina cailleti Crosse
        Tichosina floridensis, n. sp.
        Platidia clepsydra Cooper

Same locality at 213 meters
        Tichosina cubensis (Pourtalès)

Same locality at 457 meters
        Terebratulina cailleti Crosse
        Dallina floridana (Pourtalès)

U.S. GEOLOGICAL SURVEY OF WEST FLORIDA SHELF STATION

D-19   Latitude 26° 03′ N, longitude 84° 17′ W at 98 fathoms (= 179 meters) on the West Florida Shelf west-northwest of Cape Romano, Florida. Number 17 on map.
        Terebratulina cailleti Crosse
        Tichosina floridensis, n. sp.

GERDA STATION

114   Latitude 24° 02′ N, longitude 83° 02′ W at 759-869 meters, Straits of Florida, southwest of Dry Tortugas, Florida. Number 24 on map.
        Ecnomiosa gerda, n. sp.

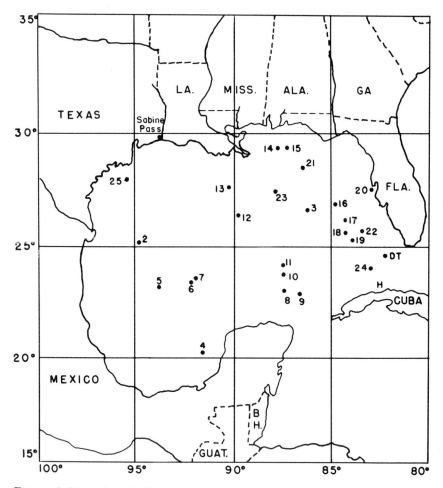

FIGURE 6. Map of the Gulf of Mexico showing dredging stations where brachiopods have been collected.

46

Figure 6. (Continued) Dredging Stations. Final figure gives depth in meters.

1. OREGON II
   Station 10962, 723 m.
2. ALAMINOS
   Station 4, 3565 m.
3. OREGON
   Station 1189, 3111 m.
4. SILVER BAY
   Station 961, 47 m.
5. ALAMINOS
   Station 4c, 3852 m.
6. ALAMINOS
   Station 5, 3843 m.
7. ALAMINOS
   Station 12, 3477 m.
8. OREGON
   Station 4574, 384 m.
9. OREGON
   Station 4570, 915 m.
10. SILVER BAY
    Station 1184, 275 m.
11. OREGON II
    Station 11133, 375 m.
12. OREGON
    Station 2574, 2414 m.
13. ALBATROSS
    Station 1408, 366 m.
14. ALBATROSS
    Station 2388, 64 m.
15. ALAMINOS
    Station 9, 732 m.
16. OREGON
    Station 955, 183 m.
17. U.S. Geological Survey
    Station D-19, 179 m.
18. BLAKE
    Station 45, 185 m.
19. OREGON
    Station 1025, 119 m.
20. Egmont Key
    and Tampa Bay
21. ALBATROSS
    Station 2400, 123 m.
22. OREGON
    Station 35, 110 m.
23. ALBATROSS
    Station 2383, 2161 m.
24. GERDA
    Station 114, 759-869 m.
25. 150 miles southwest
    of Sabine Pass, 73 m.
DT. Dry Tortugas, Florida,
    183-457 m.
H. Havana, Cuba,
   1437 m.

HEWATT COLLECTION

150 miles southwest of Sabine Pass, Texas at 40 fathoms (= 73 meters). Number 25 on map.

Argyrotheca hewatti, n. sp.

HOURGLASS STATION

M   About 92 nautical miles due west of Sanibel Island, Florida, latitude 26° 24′ N, longitude 83° 43′ W at 73 meters.

Glottidia pyramidata (Stimpson)
Platidia clepsydra Cooper

OREGON STATIONS

35   Latitude 25° 35′ N, longitude 83° 42′ W at 60 fathoms (= 110 meters) on the West Florida Shelf off Cape Sable, Florida. Number 22 on map.

Platidia clepsydra Cooper

955   Latitude 27° 08′ N, longitude 84° 53′ W at 100 fathoms (= 183 meters) on the West Florida Shelf west of Sarasota, Florida. Number 16 on map.

Terebratulina cailleti Crosse
Tichosina floridensis, n. sp.

47

1025   Latitude 25° 13′ N, longitude 83° 55′ W at 65 fathoms (= 119
        meters) on the West Florida Shelf, west of Cape Sable, north
        of Dry Tortugas, Florida. Number 19 on map.
        Terebratulina cailleti Crosse
        Tichosina floridensis, n. sp.

1189   Latitude 26° 00′ N, longitude 86° 05′ W at 1700 fathoms (= 3111
        meters) on the Mississippi Cone east of Cape Romano, Florida.
        Number 3 on map.
        Tichosina cubensis (Pourtalès)

1408   Latitude 28° 02′ N, longitude 90° 15′ W at 200 fathoms (= 366
        meters) on the Mississippi Cone south of New Orleans, Louisi-
        ana. Number 13 on map.
        Tichosina ovata, n. sp.

2574   Latitude 26° 34′ N, longitude 89° 53′ W at 1450 fathoms (=
        2414 meters) on the Mississippi Cone slightly east of south of
        New Orleans, Louisiana. Number 12 on map.
        Chlidonophora incerta (Davidson)

4570   Latitude 23° 13′ N, longitude 86° 28′ W at 500 fathoms (= 915
        meters) on the Campeche Shelf northwest of Cape San Antonio
        on the west end of Cuba. Number 9 on map.
        Terebratulina cailleti Crosse
        Platidia anomioides (Scacchi and Philippi)
        P. davidsoni (Deslongchamps)
        Ecnomiosa gerda, n. sp.

4574   Latitude 23° 13′ N, longitude 87° 50′ W at 210 fathoms (= 384
        meters) on the Campeche Shelf, northwest of Cape San Antonio
        on the west end of Cuba. Number 8 on map.
        Stenosarina oregonae, n. sp.
        Dallina floridana (Pourtalès)

OREGON II STATIONS

10962   Latitude 18° 26′ N, longitude 94° 26′ W at 395 fathoms (= 723
        meters) on the Campeche Shelf north of Coatzcoalcos, Mexico.
        Number 1 on map.
        Ecnomiosa gerda, n. sp.

11133   Latitude 24° 18′ N, longitude 87° 50′ W at 205 fathoms (= 375
        meters) on the Campeche Shelf, north of Yucatan Peninsula,
        Mexico. Number 11 on map.
        Stenosarina angustata, n. sp.
        Dallina floridana (Pourtalès)

Dry Tortugas, Florida, at 30 fathoms (= 55 meters). DT on map.

Argyrotheca lutea (Dall)

Same locality at 43 fathoms (= 79 meters).

Argyrotheca lutea (Dall)
A. rubrotincta (Dall)

SILVER BAY STATIONS

961   Latitude 20° 02′ N, longitude 91° 58′ W at 26 fathoms (= 47 meters) north of Carmen, Campeche Shelf, Mexico. Number 4 on map.

Argyrotheca sp. 4.

1184   Latitude 23° 56′ N, longitude 87° 32′ W at 150 fathoms (= 275 meters), Campeche Shelf, north of Yucatan Peninsula, Mexico. Number 10 on map.

Stenosarina angustata, n. sp.

Class INARTICULATA Huxley, 1869

Order LINGULIDA Waagen, 1885

Superfamily LINGULACEA Menke, 1828

Family Lingulidae Menke 1828.

Genus *Glottidia* Dall 1870

Glottidia pyramidata (Stimpson)

Plate 2, figure 12; plate 27, figures 1, 2

Lingula pyramidata Stimpson, 1860: 444. Brooks, 1879: 25-112.
Glottidia pyramidata (Stimpson) Dall, 1870: 158; 1873: 204.—Morse, 1902: 315-320, pls. 39-48, 50-54, 57.—Davidson, 1888: 223.—Paine, 1963: 255-280 (separate 187-213).

This species can be recognized by its small size and usually white and brownish color. Some specimens, however, are light or dark brown and some have a faint to fairly strong patch of green. Dall (1920) stated that the shell never "shows any greenish color" but specimens in the national collection from Sarasota and Tampa Bay, Florida on the Gulf of Mexico are tinted with green. Specimens sent to the National Museum of Natural History from the coast of the Carolinas and some specimens already in the collection are mostly of young shells with no green in them. Fully grown individuals seem not to be common. The fully grown adult measures about an inch (25 mm) in length (Paine, 1963: 205) with the pedicle attaining about twice the length of the shell.

No specimens of this species were taken by the research vessels of the University of Miami, but specimens in the national collection represent the entire geographic range from Chesapeake Bay to the east coast of Florida and along the west and north coasts of Florida. The species is usually found in shallow water in a sandy bottom, from tidal zone to 40 fathoms (= 73 meters).

Although Dall (1920: 269) discredits the reported occurrence of *Glottidia* at Martinique, a specimen (USNM 550736) in the National Museum was taken from Anasco Bay, on the east side of Puerto Rico, definitely showing the presence of the genus in the Caribbean. The specimen is small but appears characteristic of *G. pyramidata*. It was taken from muddy sand at 7.5 meters.

Paine (1963) has made an exhaustive study of the ecology of *G. pyramidata*. It is interesting to note that Paine reported virtual absence of *Glottidia* at Beaufort, North Carolina, where it once was very abundant. The species should be looked for in sandy areas throughout the Caribbean and Gulf regions.

Types.—Figured hypotypes: USNM 111038a, 334748a.

Glottidia audebarti (Broderip)

Plate 19, figure 1

Lingula audebardi Broderip, 1835: 143, pl. 23, fig. 14.—Sowerby (G. B.), 1847: 338, pl. 67, fig. 5.—Davidson, 1852: 377.
L. audebardii (Broderip) Küster, 1843: 15, pl. 1, figs. 10, 11.
Glottidia audebarti (Broderip), Davidson, 1888 (part): 223, pl, 28, figs. 7-9 (not 10-11).—Dall, 1920: 268.

This is a large and beautifully colored species, the pedicle and posterior being cream white to pale yellow and the anterior median half a bright green.

Locality.—P566, P572.

Types.—Figured hypotype: USNM 550536.

Order ACROTRETIDA Kuhn, 1949.

Suborder ACROTRETIDINA Kuhn, 1949

Superfamily DISCINACEA Gray, 1840

Family Discinidae Gray, 1840

Subfamily Disciniscinae Schuchert and LeVene, 1929

Genus *Discradisca* Stenzel, 1964

Discradisca Stenzel, 1964: p. 627.

This name was proposed by Stenzel (1964) as a subgenus of *Discinisca,* because of the radial costellae on the exterior. The subgenus *Discinisca* has a smooth dorsal valve. The trinomial is not used in this paper and *Discradisca* is elevated to generic rank. The genus is rare in the Caribbean but several specimens were taken by early collectors and found their way into the national collection. Unfortunately, the geographic and bathymetric data with the specimens are meager. The genus should be watched for in future dredging operations.

Discradisca antillarum (D'Orbigny)

Plate 2, figures 13-24.

Orbicula antillarum D'Orbigny, 1846 [1845, 1853]: 368, pl. 28, figs. 34-36; 1853, p. 371, pl. 1, fig. 2 (see note in references).
Discinisca antillarum D'Orbigny) Dall, 1871: 42; 1873: 201; 1920: 278.— Davidson, 1888: 204, pl. 26, figs. 31, 31a.

Unevenly subcircular in outline and conical in profile, the apex located about 1/3 the length from the posterior side; posterior slope steep, flat to slightly concave; anterior slope long and fairly steep, somewhat flattened. Apex smooth; sides and posterior marked by thread-like radii, 3 or 4 to the millimeter at the margin; anterior slope indistinctly marked by radii. Color pale brownish yellow.

Ventral valve subcircular, concave medially but gently convex marginally; marked by fine radii, 3 or 4 to the millimeter around the entire valve; pedicle region subcircular to slightly elongate oval and narrowing anteriorly; pedicle region (listrium) occupying about 1/4 the area of the valve; pedicle opening a narrow slit. Dorsal valve overlapping ventral valve slightly.

Measurements in mm	length	width	height	thickness
64335a	9.8	10.8	3.9	3.5?
64335b	10.2	10.0	3.1	1.8?
64335c	10.6	9.5	4.0	?

Localities.—Port Royal (442396), Palisadoes (442458) and Kingston (442685), all in Jamaica.

Types.—Figured hypotypes: USNM 64335a-c, 364180a.

Discussion.—Davidson (1888) regarded this as an uncertain species. It has been only poorly described and no good illustrations have been published. The few published remarks on the genus accord well with the specimens here illustrated. These specimens were originally identified by Dall who apparently regarded the species as a valid one. It is evidently a rare species but, if the few identifications of it are correct, it has a fairly wide distribution in the Caribbean region.

The national collection includes two lots of *Discradisca* from Brazil referred by Dall to *D. antillarum*. These are of about the same size as those described above, but the shell is completely radiate and the apex is more nearly central than that feature of the Jamaica specimens. This probably represents another species, but too few specimens are at hand to establish the species. It is another record of brachiopods from the Brazilian coast, which is a critical area because along this coast should come a mingling of elements from the north with those of the south. The mentioned and figured specimens (USNM 364180a-c) are from Sao Francisco, Santa Catarina, Brazil.

Genus Pelagodiscus Dall, 1908

Pelagodiscus atlanticus (W. King)

Plate 1, figures 8-13

Discina atlantica W. King, 1868: 170.–Jeffreys, 1876: 252.–Davidson, 1880: 62, pl, 4, figs. 17, 18.

?*Discinisca atlantica* (W. King) Dall, 1873: 261.—Davidson, 1888: 200, pl. 26, figs. 18-22.

Pelagodiscus atlanticus (W. King) Dall, 1908: 440.—Thomson, 1918: 38, 40, 50.—Dall, 1920: 280.—Thomson, 1927: 130.—Helmcke, 1940: 230 (extensive synonymy).—Hertlein and Grant, 1944: 21 (extensive bibliography).—Zezina, 1965: 345-358.

Pelagodiscus atlanticus (W. King) is easily recognized by its circular outline, conical profile, yellow to brown color and the fringe of long, barbed setae around the margin. The ventral valve is concave and the pedicle large. This brachiopod is widespread and is known from the deeps of all the oceans. It is reported from 6160 meters, one of the greatest depths from which a brachiopod has been taken (Zezina, 1965: 5).

The national collection has no specimens of this species from the Gulf of Mexico but its presence there was reported by Dr. Willis E. Pequegnat (letter of February 20, 1967). It has been taken from several localities in the Atlantic and Caribbean, and some specimens from off Georgia are illustrated herein.

Localities.—PILLSBURY Stations P120, P338, P976, P1181.

Types.—Figured hypotypes: USNM 550729a, 550739a, b, 550730.

Suborder CRANIIDINA Waagen 1895

Superfamily CRANIACEA Menke, 1828

Family Craniidae Menke, 1828

Genus *Crania* Retzius, 1781

Crania pourtalesi Dall

Plate 2, figures 5-13, plate 3, fig. 18

Crania anomala pourtalesii Dall, 1871: 35, pl. 1, figs. 7a, b.

This is a rare and poorly known species. It is flatly conical, subcircular or faintly elliptical in outline. The type specimen measures 8 mm. wide by 7 mm. long and about 2 mm. in height. The color is light to dark brown. The apex is rounded and located about 1/3 the valve length from the posterior margin. The posterior slope is gently concave and steep; the anterior slope is gently convex. The surface is very irregular as a consequence of its fixation to a rough surface. Strong concentric lines are the only markings; no spines or radii were seen.

The ventral valve of the type is badly fractured but it shows the pits made by the internal oblique muscles. A second specimen (111023) is well preserved and many details are available. The specimen is transverse. The oblique internal muscles are attached in two deep pits in a dorso-central thickening. The posterior and anterior adductor scars are rather small and not strongly defined. The vascula lateralia are thick and compressed. The margin of the valve is strongly thickened and finely granulose.

The dorsal valve is less thickened than the ventral valve and is a rather delicate shell. The adductor muscle scars are of moderate size and located in the posterior half. The protractor muscles occupy an elevated, elongated thickening supported by a short septum.

The pallial trunks are not impressed.

Types.—Holotype: USNM 111022. Figured hypotypes: USNM 111023, 549448a.

Localities.—The type specimen (111022) comes from the Sambos, Florida, in 114 fathoms according to the label (209 meters) on a rocky bottom (text gives 116 fathoms for this specimen); a second specimen (111024) from Sand Key, Florida, is given as 114 fathoms on the label but 105 fathoms in the text. A third specimen, very similar to the type specimen, is said to have come from 226 fathoms (=334 meters) off Cuba. A dorsal valve from Johnson-Smithsonian Station 52 bears two attached *Lacazella*.

Discussion.—This species was originally referred to *Crania anomala* (Müller), which is common in the waters around the British Isles, Norway and Portugal. This is a larger and thicker-shelled species than the Caribbean one. The European species does not have the muscle areas so strongly thickened as those of *C. pourtalesi*. This thickening is a trend in the direction of *Craniscus* in which the dorsal muscles are attached to a Y-shaped platform.

Crania aff. *C. pourtalesi* Dall

Plate 28, figures 1-3

Crania anomala var. *pourtalesi* Dall, 1871: 35, pl. 1, figs. 7a, b.
Crania pourtalesi Dall, 1920: 273.

The national collection contains only a single specimen of this genus from the Gulf of Mexico. This specimen comes from the Campeche Bank (Shelf). It is pale yellowish, a depressed cone with apex near the posterior margin. The posterior slope is thus very short but steep. The exterior is marked by fine lines of growth and some irregularities produced by its growth on an uneven surface. The interior has the usual large posterior adductor scars but the median muscles have thickened bases and a median ridge is fairly strongly developed. The specimen is nearly circular and measures 8.5 mm in length and width.

Locality.—Latitude 23° 18' N, longitude 87° 02' W, Campeche Bank (Shelf), north of Cabo Catoche, Mexico, at 200 fathoms (= 366 meters).

Types.—Figured specimen: USNM 61225.

Discussion.—The specimen figured herein differs from the type specimen of *C. pourtalesi* Dall taken off the Sambos reefs, Florida. The type is a much higher cone, squarish in outline and dark brown in color. The apex is a little posterior of the center and has a long, steep posterior slope. The interiors of the two are similar as might be expected, but the exteriors, even though cemented brachiopods are variable, are so different as to suggest separate species.

Crania sp.

Crania sp. Tunnell, 1972: 553.

Tunnel mentions a specimen of *Crania* taken from Seven and One-Half Fathom Reef, latitude 26° 51' N, longitude 91° 18' W, southern Texas. The specimen measures 12 mm in long diameter, 8 mm in short diameter and 4 mm in height. Its special character is the presence of radial lines on its surface, unlike any modern *Crania* from Caribbean or Atlantic waters. Some Tertiary (Eocene) Cranias are however, so marked. The specimen is deposited in the invertebrate collection of the Texas A. & I. University.

54

Class ARTICULATA Huxley, 1869

Order RHYNCHONELLIDA Kuhn, 1949

Superfamily RHYNCHONELLACEA Gray, 1848

Family Cryptoporidae Muir-Wood, 1955

Genus *Cryptopora* Jeffreys, 1869

Cryptopora rectimarginata Cooper

Plate 2, figures 1-4; plate 5, figures 1-8; plate 27, figures 7-14

Cryptopora gnomon Cooper, 1954 (not Jeffreys): 364.
C. rectimarginata Cooper, 1959: 20, pl. 1, figs. 15-19, pl. 2, figs. 1-11.

Rhynchonellid brachiopods are rare in the Caribbean and Gulf of Mexico. *Cryptopora* is the only one now known in these regions and is so small that it may be easily overlooked. Although small it is very distinctive. It has a silvery, lustrous shell having a coarse mosaic formed by the shell fibers. These may be mistaken for punctae. It is roundly triangular in outline and lenticular in profile with a faintly uniplicate to rectimarginate anterior commissure. The delthyrium in this species is margined by narrow, elevated, often winged deltidial plates.

The interior of *Cryptopora* is as distinctive as its exterior. Inside the ventral valve there are strong dental plates, but other details, such as muscle or pallial marks, are usually faint. The interior of the dorsal valve is readily recognized by its long crural plates with their hand-like expansion, a type of rhynchonellid crus known as "maniculifer." The median septum occupies the center of the valve where it is strongly elevated and forms a truncated crest. The anterior edge is steep but the dorsal edge tapers to the posterior and usually does not reach the notothyrial chamber. A small cardinal process is usually present.

Locality.—ALBATROSS Station 2400. EOLIS 340. GERDA G289, G678, G691.

Types.—Holotype: USNM 274134a. Figured paratypes: USNM 274143c, d. Figured hypotypes: USNM 323891, 550731, 550732.

Discussion.—This species is rare in the Gulf of Mexico and only one lot consisting of two specimens is present in the national collection. It is better known in the Straits of Florida where it was collected in some abundance by J. B. Henderson off Fowey Light, Florida, in water ranging from 137 meters to 382 meters. The specimens collected by the R/V GERDA are from deeper water, down to 604 meters.

Cryptopora has a worldwide distribution and occurs in shallow as well as very deep water. In the North Atlantic, *C. gnomon* (Jeffreys) occupies water as deep as 2900 meters. The genus is found as a fossil in the Oligocene and Miocene of Cuba, and Eocene of the Gulf Coast.

Cryptopora externally looks very much like the earliest stages of *Dallina floridana* (Pourtalès) which is triangular and has flaring deltidial plates at this stage (see plate 4, figures 1-4). The two can be separated readily by examination of the shell structure, without pores in *Cryptopora* but well provided with them in the punctate *Dallina*.

Order TEREBRATULIDA Waagen, 1883

Suborder TEREBRATULIDINA Waagen, 1883

Superfamily TEREBRATULACEA Gray, 1840

Family Dyscoliidae Fischer and Oehlert, 1891

Genus *Dyscolia* Fischer and Oehlert, 1890

Dyscolia wyvillei (Davidson)

Plate 15, figures 1-10

Terebratulina wyvillii Davidson, 1878: 436; 1880: 32, pl. 1, figs. 1, 2; 1886: 32, pl. 3, figs. 1-3.

Dyscolia wyvillei (Davidson), Fischer and Oehlert, 1891: 23-29, pl. 6.

Very large, elongate oval in outline, sides well rounded, anterior margin rounded. Posterolateral margins forming an angle of 80°. Foramen large, moderately labiate, transversely and roundly elliptical, permesothyridid. Surface marked by strong incremental lines of growth and low interrupted radial lines. Punctae densely crowded.

Ventral valve moderately convex in profile, maximum convexity in the posterior third; anterior profile forming a narrowly convex, steep-sided dome. Symphytium thick, concentrically marked and concave.

Dorsal valve with variable convexity, fairly flattish in the young, only moderately but evenly convex in the adult; anterior profile a broad, shallow dome with long gentle slopes; umbonal region narrowly swollen; umbo-lateral slopes steep and forming posterolateral flattened or concave areas on each side of the valve posterior; lateral margins often abruptly deflected by a change in growth direction of almost 90°.

Pedicle valve interior with strong but narrow teeth; pedicle collar short but excavated for its full circle; muscle scars not impressed strongly enough to reveal their pattern.

Dorsal valve with small, transverse cardinal process; socket ridges short, not strongly elevated; outer hinge plates moderately concave, small and short, confined. Loop not preserved in the PILLSBURY specimens. Adductor scars elongate, somewhat tear-shaped and widely separated.

Measurements in mm	length	dorsal valve length	width	thickness	apical angle
550621a	57.6	?	46.8*	17.6	?
b	42.7	?	37.0	16.0	80°
c	?	43 plus	47.0*	15.7?	?
d	?	41 plus	44.0*	16.0?	?

*Based on half measure.

Locality.—P1262.

Types.—Figured hypotypes: USNM 549271, 550621b, c, e.

56

Discussion.—This is the largest brachiopod known in the Caribbean region. It was reported by Davidson in the CHALLENGER report but only one specimen was found. The PILLSBURY specimens consist of two ventral and four dorsal valves, all dead shells and all imperfect but very important nevertheless. They authenticate Davidson's report of this species in the Caribbean and afford a new location that indicates a fairly wide distribution.

Although all of the specimens are imperfect, each has important characters. These added together corroborate Davidson's description and that of Fischer and Oehlert based on specimens taken off the northwest coast of Africa.

Exterior as well as interior characters indicate the relationship of this PILLSBURY material to *Dyscolia*. Aside from the primitive loop the important generic characters of *Dyscolia* are the large foramen and thick, visible symphytium; the prominent umbonal region of the dorsal valve and the depressed posterolateral regions; the subdued but well marked radial ornament and finally the peculiar growth form that results in an abrupt inward deflection of the anterior margin. In a complete specimen this produces a peculiar flattening of the anterior and lateral margins and in some specimens a concave marginal zone around the margins.

Although the loop is missing from the PILLSBURY specimens, the hinge plates are like those of *Dyscolia* and the cardinal process is of the same type. This is a feature not shared by *Terebratulina,* to which this species was first assigned.

Davidson's drawing of the dorsal interior of this species is undoubtedly incorrect. He does not say whether or not the loop was preserved in his single specimen but the suspicion is that it was not. He has depicted (Davidson, 1880, pl. 1, fig. 2) a dorsal valve with a loop having the complete ring of *Terebratulina.* The loop of *Dyscolia* is very simple, with short blunt crural processes that are not directed medially and could not meet. The loop of a specimen from northwest Africa is introduced on plate 15 to show the true character of this important structure.

Dyscolia has a wide and interesting distribution. It occurs in the Indian Ocean off the Maldive Islands, where it lives at a depth of 787-1135 meters (Muir-Wood, 1959). It occurs off the Azores, north of Spain, and off the coast of Morocco in depths from 712-1134 meters. Davidson's specimen was taken off Culebra Island, northwest of St. Thomas, at 1263 meters. The genus recently has been found off the coast of Argentina and in the Antarctic. It is known as a fossil in the Pliocene of Sicily and the Tertiary of Chatham Islands east of New Zealand.

Family Terebratulidae Gray, 1840

Subfamily Terebratulinae Gray, 1840

The Subfamily Terebratulinae comprises a large group of brachiopods with a long geological history. They are mostly without conspicuous radial ornament, and have short loops of fairly uniform aspect. The Terebratulinid roots in the Paleozoic are uncertain, but the subfamily was abundant in the Triassic and was a large element of the brachiopod population in the Jurassic. In the Cretaceous it was still an abundant group with numerous genera mostly separable on features of the loop and cardinal process. After Cretaceous time, terebratulinids declined in number of genera but the genera that were represented were fairly abundant locally. At the present time, the subfamily is represented by four genera only. *Liothyrella* is restricted to the Southern Hemisphere; *Abyssothyris* is restricted to the abyss. *Gryphus,* the type of which is best known from the Mediterranean, is widely but rather uncritically identified in all oceans and is in great need of revision. Its loop is very much like that of *Dallithyris* from the Indian Ocean.

The Caribbean and Gulf of Mexico have a great concentration and variety of Terebratulinae. Although one of the commonest Caribbean species has been referred to *Dallithyris,* close examination of the cardinalia of the Caribbean forms indicates that this genus is not now represented there. Generic differentiation on the basis of cardinalia is difficult in these generalized brachiopods and any scheme of splitting is likely to be subject to criticism by the more conservative taxonomists.

Of all the morphological characters to be seen in these difficult brachiopods, only the cardinalia offer valid criteria for differentiation. Exterior characters seem inconstant. The sulcate character of the anterior commissure has been used to separate *Abyssothyris.* The shallower water terebratulinids usually have an anterior commissure ranging from rectimarginate to strongly uniplicate, but it is commonly variable within these limits in a single population. Size of foramen is a helpful but not constant character.

The cardinalia thus offer the best means for differentiation, yet these are difficult and variable. The cardinal process is usually so poorly developed as to be negligible. The outer hinge plates and crura, together with the general form of the loop, are the most helpful features. The transverse ribbon is also a useful character. These structures combined with general form of the exterior are the features that provide meaningful distinctions. For example, the case of "*Gryphus*" *cubensis* (Pourtalès) will illustrate. Very largely on the basis of its plump triangular form, Muir-Wood (1959) assigned it to *Dallithyris.* Its loop is narrow like that of *Dallithyris* but the detail of its hinge plates is quite different; consequently, as explained below, "*G.*" *cubensis* is here assigned to a new genus having loop characters similar to those of other species in the Caribbean.

Genus *Abyssothyris* Thomson, 1927

Abyssothyris atlantica Cooper, new species

Plate 20, figures 1-10

Small, longitudinally elliptical in outline, with well rounded sides but narrowly rounded anterior margin; posterolateral margins forming an angle of 81°. Lateral commissure gently curved in a dorsad direction; anterior commissure broadly sulcate. Beak suberect, truncated, labiate; foramen moderately large, submesothyridid; symphytium visible. Valves strongly inequivalve. Color translucent but with a dark brown epidermis. Surface marked by concentric growth lines. Punctae numbering 188 per square millimeter.

Ventral valve strongly convex in lateral profile, with the maximum convexity in the posterior part; anterior profile strongly and broadly domed, with precipitous sides. Beak and umbo moderately elongated, narrow, with convex and rounded sides.

Dorsal valve with less than half the depth of the ventral valve and with a strongly convex lateral profile having the maximum convexity in the posterior half and the anterior flattened; anterior profile a slightly convex lid over the ventral valve. Tongue broadly rounded but short. Anterior half of dorsal valve flattened to form a poorly defined sulcus.

Pedicle valve interior with narrow, elongated teeth and short anteriorly excavated pedicle collar. Pedicle thin but fairly long. Dorsal valve interior with low, thin socket ridges; loop measuring one-third the valve length, inner hinge plates narrow and concave; no marginal rims; descending branches attached to inner hinge plate broad and forming somewhat rounded crural processes; transverse ribbon elongated anteriorly, narrowly rounded and with a low but sharp median angulation; no anterolateral angulation of the loop. Lophophore narrow, tightly coiled, short, with short lateral branches.

Measurements in mm	length	dorsal valve length	width	thickness	apical angle
550592a (holotype)	11.0	10.0	7.9	7.0	81°
550592b	10.1	8.8	7.8	?	80°

Locality.—Hydro Station 5809.

Diagnosis.—Very small *Abyssothyris* with broadly sulcate anterior margin.

Types.—Holotype: USNM 550592a. Figured paratypes: USNM 550592b, c.

Discussion.—Discovery of a North Atlantic species of this abyssal genus is of considerable interest. *Abyssothyris atlantica* differs from *A. wyvillei* (Davidson), the type of the genus, in being much smaller, more elongate, and in having a much broader fold of the anterior commissure. Inside the dorsal valve, the type-species has a laterally angulated and wider loop than that of the Atlantic species. A species of *Abyssothyris* from the late Tertiary of Fiji is too much like the type-species to need comparison with the Atlantic form. Specimens in the national collection from off the Galapagos Islands are larger and much wider than *A. atlantica*. This is the first report of *Abyssothyris* in the northern Atlantic; it is best known from the Pacific and Antarctic Oceans.

Abyssothyris? parva Cooper, new species

Plate 18, figures 1-7

Very small for the genus, subtriangular in outline, with the greatest width anterior to midvalve; sides rounded; anterior margin nearly straight and producing narrowly rounded anterolateral extremities. Lateral commissure straight; anterior commissure rectimarginate. Posterolateral margins forming an angle of about 85°. Beak erect, truncated; foramen small, not labiate, mesothyridid; symphytium visible. Color white; marked by concentric growth lines. Punctae per square millimeter, 175-188.

Ventral valve much deeper than the dorsal valve, moderately convex in lateral profile with the umbonal region the most convex, anterior profile broadly and gently convex; anterior region convex; flanks rounded.

Dorsal valve gently convex in lateral profile with the maximum convexity at about midvalve; anterior profile broadly and gently domed and with short lateral slopes. Anterior slope short but convex.

Ventral interior with small narrow teeth; pedicle collar fairly long and anteriorly excavated. Dorsal valve with stout socket ridges; loop in length equal to one-third the dorsal valve length; loop fairly narrow, parallel-sided. Outer hinge plates narrowly elongated, flatly concave, without marginal elevations; crural base narrow and supporting a bluntly pointed crural process; transverse ribbon moderately broad, rounded, without anterolateral angles and not folded medially. Lophophore short, without long lateral branches.

Measurements in mm	length	dorsal valve length	width	thickness	apical angle
550593a	8.1	6.8	7.2	4.8	85°

Locality.—ATLANTIS Stations A-266-2, A-266-4.

Diagnosis.—Very small with triangular outline.

Types.—Holotype: USNM 550593a. Figured paratype: USNM 550593b.

Discussion.—This little shell suggests "*Gryphus*" *minor* (Philippi) and "*G.*" *affinis* (Calcara) but is smaller than both and has a different loop. "*Gryphus*" *minor,* beside being larger is more convex with more rounded anterior margin than that of *T. parva.* "*Gryphus*" *arcticus* (Friele) and "*G.*" *davidsoni* (A. Adams) are small species that are differently shaped than the specimens from off Georgia.

This species is tentatively assigned to *Abyssothyris* even though its anterior commissure is rectimarginate. The loop is not like that of *Tichosina* but its rounded form and lack of pointed anterolateral extremities suggest *Abyssothyris.* The loop of *Abyssothyris* is not always with rounded anterolateral extremities, because Muir-Wood (1960) figures the type with pointed extremities. Another unusual feature in *A.? parva* is the deltidial plates which show a suture line where joined. Most specimens of *Ticho-sina* have a symphytium rather than a deltidium. This loop and the lightly coiled lophophore it supports are most like those of *Abyssothyris elongata* Cooper from the California Abyssal Plain off Los Angeles (Cooper, 1972).

Abyssothyris sp. 1

Plate 11, figures 17-19

Gryphus moseleyi Dall (not Davidson), 1920: 318.

Small, nearly circular in outline, very thin-shelled and delicate; ventral valve much deeper and more convex than the dorsal valve. Anterior commissure perceptibly sulcate. Interior with broken loop which appears to have rounded anterolateral extremities. Punctae numbering 80 per square millimeter. Measurements in mm: length 14.7; dorsal valve length 13.0; width 14.0; thickness 8.0; apical angle 94°.

Locality.—ALBATROSS Station 2035.

Types.—USNM 110853.

Discussion.—This specimen was originally assocated with *T. martinicensis* by Dall (1920: 318) which in turn had been mis-identified as *Gryphus moseleyi* (Davidson) by T. Davidson. The specimen is placed here because of its sulcate anterior commissure and the nature of the loop. It is from 2492 meters, a depth well within that normal for the genus. More specimens however, are needed for confirmation of this determination.

Abyssothyris ? sp. 2

Brief mention is made of a small specimen from P1138 that comes from very deep water and is associated with *Chlidonophora*. It resembles *A.* ? *parva,* n. sp., but is much rounder in outline. Its apical angle is 98° whereas that of *A.* ? *parva* is only 85°. The shell is very thin and transparent, as is usual in deep sea brachiopods. Unfortunately the loop is not preserved but the hinge plates are similar to those of *Abyssothyris*. The anterior commissure is rectimarginate, which may indicate a young individual. The specimen measures 7 mm in length and 6 mm in width. Punctae number about 156 per square millimeter but the shell is poorly preserved and the punctae difficult to see.

Types.—Described specimen: USNM 550615.

Tichosina, new genus

Small to fairly large for a brachiopod, usually elongate, oval in outline with the valves of unequal depth, the ventral valve usually having the greater depth. Color usually white but some shells rarely yellow to salmon-color. Beak narrow, foramen usually small, labiate and generally permesothyridid. Anterior commissure variable from rectimarginate to strongly uniplicate. Loop narrow, with moderately developed outer hinge plates but with broad, flattish crural bases extending to the apex and walling off the outer hinge plates. Outer hinge plates attached to dorsal edge of crural base distally; crural processes blunt, anterior in position and generally overhanging the posterior of the broad, medially folded transverse ribbon.

Type-species.—Tichosina floridensis Cooper, n. sp.

*Diagnosis.—*Terebratulinae with a flat bladed crural base and crural process forming a wall along the inside edge of the outer hinge plate.

*Derivation. —*Greek *teichos* = wall.

*Comparison and discussion.—*Generic separation of the multitude of species of the Terebratulidae is fraught with difficulties. There is considerable homeomorphy of exterior details but the loop seems to offer the best possibilities for generic separation. The loop of *Tichosina* is unusual in the great development of the crural base which is extended anteriorly to join the crural process, thus making a long flat blade. The two sides of the loop are nearly parallel, close together and bounded by a broad transverse band. Superficially this loop is suggestive of *Dallithyris* and at least one of the species assigned to this genus by Muir-Wood really belongs to *Tichosina.* This is *T. cubensis* (Pourtalès), which is a homeomorph of *Dallithyris* but differs in loop characters. This species is intermediate in generic characters between *Dallithyris* and *Stenosarina.*

The loop of *Dallithyris* differs in several important respects from that of *Tichosina.* The outer hinge plates of *Dallithyris* are broad and rather flat, tapering anteriorly to a narrow, short band to which are attached the expanded crural processes and transverse ribbon. The crural bases are not elevated blades and do not form a wall along the inside edge of the outer hinge plate. Actually, the *Dallithyris* loop is almost identical to that of *Gryphus* that has the anterior part of the loop attached to the outer hinge plate by rounded, narrow bands. Details other than the character of the loop, however, separate *Gryphus* from *Dallithyris.*

The loop of *Tichosina* differs from that of *Terebratula* and *Liothyrella* in being narrower, with a much stouter and broader (in an anterior-posterior direction) transverse ribbon. The loop of the other two is wide anteriorly (but narrow in the longitudinal direction) and gives the impression of being triangular. It is also usually more delicate than that of *Tichosina.*

Tichosina abrupta Cooper, new species

Plate 10, figures 1-10

Small for the genus, narrowly oval in outline with rounded sides and anterior margin. Posterolateral extremities forming an angle of 70° to 83°. Ventral valve deeper than the dorsal valve. Lateral commissure nearly straight; anterior commissure faintly uniplicate. Beak short, truncated, suberect; foramen large, labiate, permesothyridid. Color white; surface marked by concentric lines and varices of growth.

Ventral valve strongly convex in lateral profile, with the maximum convexity in the posterior half; anterior half gently convex and forming a steep slope; anterior profile a narrow dome with steep sides. Umbonal region narrowly swollen.

Dorsal valve evenly and moderately convex in lateral profile with the maximum convexity at about midvalve; anterior profile a moderately convex dome but not so elevated as that of the ventral valve. Umbonal and median regions swollen.

Ventral valve with small teeth and short anteriorly excavated pedicle collar. Dorsal valve with small, transverse cardinal process; loop measuring in length about one-quarter the valve length; outer hinge plates long and deeply concave; crural bases forming a marginal rim to the hinge plate; crural processes attached at end of hinge plate, flat and bluntly pointed; transverse ribbon located at end of crural process, with a broad ribbon and slight median fold.

Measurements in mm	length	dorsal valve length	width	thickness	apical angle
336848	14.5	12.0	11.1	10.8	70°
550599 (holotype)	15.8	13.7	12.6	10.0	83°

Localities.—EOLIS Stations 1 and 31.

Types.—Holotype: USNM 550599. Figured paratype: USNM 336848.

Diagnosis.—Small, obese *Tichosina* with short loop.

Discussion.—This species suggests *T. bahamiensis,* new species, and *T. obesa,* new species, in size and general configuration but its short, wide loop with the long outer hinge plates is entirely unlike the other two. The loop of "*Gryphus*" *davidsoni* (A. Adams) is similar but in that species the part of the loop anterior to the crural processes is very narrow but the hinge plates are large. All specimens of *T. abrupta* are too dense to permit a count of the punctae.

Tichosina bahamiensis Cooper, new species

Plate 8, figures 10-26

Small for the genus, longer than wide and oval in outline with the maximum width at or near midwidth; sides gently rounded; anterior margin gently rounded; lateral commissure straight; anterior commissure rectimarginate. Posterolateral margins forming an angle of 70°-90°. Beak narrow, suberect, moderately extended; foramen small, labiate, permesothyridid; symphytium visible. Surface marked by concentric lines of growth; color yellow to lustrous white and translucent. Punctae 116 per square millimeter.

Ventral valve deeper than the dorsal valve and moderately convex in lateral profile; anterior profile strongly and evenly domed and with short steep sides. Umbonal region narrowly convex; anterior slope steep.

63

Dorsal valve gently convex in lateral profile and with the maximum convexity slightly posterior of midvalve; anterior profile a broad gently convex dome; umbonal region not strongly swollen. Anterior slope gentle.

Ventral valve interior with strongly impressed muscle and pallial marks; teeth small; pedicle collar short and anteriorly excavated.

Dorsal valve with narrow, parallel-sided loop extending one-third the valve length into the interior; socket ridges stout; outer hinge plates short, concave, margined by the elevated crural bases; crura long, broad and flat; crural processes anterior, bluntly pointed and overhanging the transverse ribbon which is very broad and with a strong median fold. Lophophore with long lateral branches.

Measurements in mm	length	dorsal valve length	width	thickness	apical angle
87378a (holotype)	15.2	12.9	11.3	9.2	70°
87378b	14.7	13.0	12.5	9.6	80°
87378c	14.6	12.2	11.5	8.5	71°
87378d	13.1	11.7	10.4	7.5	78°
87378e	12.3	10.8	10.7	7.7	82°

Locality.—G1177, G688. ALBATROSS Station 2655.

Diagnosis.—Small, elongate oval *Tichosina* with narrow, broad-ribboned loop and lophophore with long lateral branches.

Types.—Holotype: USNM 87378a; figured paratypes: 87378b, c, f, g. Unfigured paratypes: USNM 87378d, e.

Discussion.—Of the described smaller species this one strongly resembles "*Gryphus*" *affinis* (Calcara) but the European species is more robust, rounder, with thicker shell, more convex valves, more strongly incurved beak with larger foramen and a wider loop with narrower transverse ribbon.

Tichosina ? bartletti (Dall)

Plate 3, figures 27, 28; plate 10, figures 11-17

Terebratula bartletti Dall, 1882: 885; 1886: 200, pl. 6, figs. 4a-c.—Davidson, 1886: 14, pl. 1, figs. 20, 21.
Gryphus bartletti (Dall), 1920: 314.

Large, elongate oval in outline, with the posterior narrowing conspicuously; maximum width anterior to midvalve; sides rounded, anterior margin broadly rounded; posterolateral margins forming an angle of 75° to 85°. Lateral commissure curved anteriorly. Anterior commissure strongly uniplicate in the adult, rectimarginate in the young. Beak not greatly extended, truncated and with a moderately large, strongly labiate and permesothyridid foramen. Surface marked by concentric growth lines and fine, obscure radial lines especially on the flanks. Punctae numbering 77 (in the type) to 119 (USNM 549393) per square millimeter.

Ventral valve strongly and evenly convex with maximum convexity at about midvalve; anterior profile moderately but broadly domed and with moderate lateral slopes; median region fairly strongly swollen, the swelling extending and narrowing to the beak and also extending onto the long tongue.

Dorsal valve having about the same depth as the ventral valve and a fairly strongly convex lateral profile, the maximum curvature located posterior to midvalve and the anterior somewhat flattened. Anterior profile strongly and somewhat narrowly domed and with long steep sides; umbonal and median regions swollen, the swelling continuing anteriorly onto the fold; flanks rounded and steep; fold occupying about one-third the valve length and about two-thirds the valve width. Fold low and demarcated by narrowly rounded marginal plications.

Ventral valve interior with short elevated pedicle collar, and small elongated teeth. Pallial trunks lightly impressed; muscle area narrow, a little less than one-third the valve length. Dorsal valve with loop narrow, and nearly parallel-sided, measuring in length about one-fourth the valve length; outer hinge plates fairly long and deeply concave; crural bases forming a high margin to the outer hinge plates; crura broad and flattened; transverse ribbon broad (longitudinally) and with a gentle median fold.

Measurements in mm	length	dorsal valve length	width	thickness	apical angle
110852 (holotype)	42.3	38.6	32.4	26.7	81°
64257a	36.5	33.2	27.6	24.0	81°
550613	41.9	?	31.4	?	81°

Localities. G1012. P657, P907. BLAKE 157. Johnson-Smithsonian 102.

Types.—Holotype USNM 110852. Figured hypotype USNM 549393a.

Diagnosis.—Very large, strongly uniplicate *Tichosina?*.

Discussion.—This is the largest member of the subfamily Terebratulinae so far found in the Caribbean region and can usually be identified by its strong uniplication and the posterior narrowing. The type specimen is of a pale salmon color and another specimen (USNM 550611) is still paler salmon color, but most of the other material identified with this species is translucent white. The salmon color may be due to bottom conditions in which the dead shell lay and is probably not the normal color. Of the few lots of this uncommon species in the national collection, all but one are from the Lesser Antilles. The one is from south of the east end of Cuba, off Santiago de Cuba. Two other large species of *Tichosina* occur, one of them from off Nicaragua and the other south of Great Inagua. Although large and uniplicate, these two will not be confused with typical *T. ? bartletti* as they are much rounder.

65

The generic assignment is queried because this species is an aberrant form, deviating from the more average representatives of *Tichosina* in having a small, strongly labiate foramen, stronger folding than usual, and with outer hinge plates longer than the average. The loop is more like that of some Eocene species such as *"Terebratula" wilmingtonensis* (Lyell and Sowerby) from North Carolina.

Tichosina bartschi (Cooper)

Plate 11, figure 16

Gryphus bartschi Cooper, 1934: 1, pl. 1, figs. 1-8.

This species is introduced for comparison with *T. subtriangulata,* n. sp., which resembles *T. bartschi.* The two have similar coloring and size but *T. bartschi* narrows anteriorly while *T. subtriangulata* widens anteriorly and has its greatest width in the anterior region. Punctae per square mm: 64.

Tichosina bartschi seems to be a very rare species because none of the collections of the GERDA and PILLSBURY have contained it. The species was taken on the north or Atlantic side of the Virgin Islands.

Types.—Holotype: USNM 431002.

Locality.—Smithsonian-Johnson Expedition Station 102.

Tichosina bullisi Cooper, new species

Plate 6, figures 1-8

Large, translucent white, rounded oval in outline with the greatest width at midvalve; sides well rounded, anterior margin somewhat narrowly rounded; posterolateral margins forming an angle of 84°. Lateral commissure slightly curved anteriorly; anterior commissure with a gentle fold toward the dorsal side. Beak not greatly labiate, permesothyridid. Surface marked by concentric lines only; punctae 84 per square millimeter.

Ventral valve slightly deeper than the dorsal valve and with strongly and evenly convex lateral profile; anterior profile broadly and moderately domed and with short rounded slopes; median region swollen; anterior slope gently convex and extended anteriorly to form a short, truncated tongue.

Dorsal valve moderately and evenly convex in lateral profile; anterior profile stronger than that of the ventral valve, somewhat narrowly domed with flattened and long slopes; fold originating in the anterior third, broad and gently convex, only slightly elevated. Median region narrowly swollen, the swelling flattening toward the fold.

Teeth small; pedicle collar short with elevated rim, muscle field moderately impressed; diductors somewhat flabellate; pallial trunks not well developed.

Dorsal valve interior with semielliptical, small cardinal process; socket ridges moderately strong; outer hinge plates fairly wide, concave and long, margined by a strongly elevated crural base; crura short and broad, crural processes sharply pointed and overhanging the wide and broadly folded transverse ribbon. Loop in length about one-third the length of the brachial valve. Body wall not strongly spiculate.

Measurements in mm	length	dorsal valve length	width	thickness	apical angle
550609a	34.3	30.7	29.2	20.6 ·	84°
550609b	36.2	32.8	30.9	20.0	84°
550609c	36.5	32.0	31.4	21.3	90°

Locality.—OREGON Stations 3608, 6423.

Diagnosis.—Roundly oval, large *Tichosina* resembling *T. bartletti* (Dall) but having a shorter and lower fold.

Types.—Holotype USNM 550609b; unfigured paratypes: USNM 550609a, c.

Discussion.—This species approaches *T. ? bartletti* (Dall) in size and general appearance but it is a more roundly oval shell, with a larger foramen which is much less labiate than that of Dall's species. The folding of *T. bartletti* is much more pronounced than that of *T. bullisi,* with the result that the former has a much longer tongue in the ventral valve.

Tichosina bartletti in its typical form appears to be concentrated in the eastern Caribbean and *T. bullisi* comes from the extreme west side. This species is named in honor of Harvey Bullis who collected the specimens and has added greatly to our knowledge of the Caribbean and West Indian brachiopods.

Tichosina cubensis (Pourtalès)

Plate 22, figures 1-8; plate 27, figures 15-22

Terebratula cubensis Pourtalès, 1867: 109.—Dall, 1871: 3, pl. 1, figs. 2, 8-15.—
  Davidson, 1880, p. 28, pl. 2, figs. 10, 11.
T. vitrea var. *sphenoidea* Jeffreys (not Philippi), 1878: 404, pl. 22, fig. 6.
Lyothyris sphenoidea Davidson (part) not Philippi, 1886: 12, pl. 2, figs. 19a,
  b, 21, 22.
Gryphus cubensis (Pourtalès) Dall, 1920: 315.—Cooper, 1954: 364.

This is a large species distinguished by its plump valves and rounded, triangular outline. Its anterior commissure is rectimarginate to broadly and slightly uniplicate; the lateral commissure is curved anteriorly toward the ventral valve. The foramen is large. Inside the dorsal valve the loop is narrow and has a broad transverse band which is strongly folded medially. The outer hinge plates are bordered by elevated crural bases.

Tichosina cubensis externally resembles *Dallithyris* in its triangular outline and usually S-shaped lateral commissure. Its loop, however, is margined by elevated crural bases, although it is narrow and suggestive of *Dallithyris.* The latter does not have these plates margined by the crural bases. The external form is also similar to that of *Stenosarina,* n. gen., described below but it is usually more protuberant at the beak and has a much larger foramen. The loop of *T. cubensis* is variable, some being rather narrow and with rounded anterolateral extremities suggestive of the loop of *Stenosarina*. More commonly the loop is moderately wide, usually wider than that of *Stenosarina*, and the lateral extremities are not tapering. This species is thus rather intermediate between *Dallithyris* and *Stenosarina* having the shape of the former and the loop of the latter. Because of the large foramen, suberect beak and tichosinid loop, it is placed in *Tichosina*.

Tichosina cubensis is commonest in the Straits of Florida and has been identified as far north as the coast of Georgia. Dall lists this species from several places in the Caribbean and off Cuba but most of these specimens are small, somewhat narrow forms that do not belong to the species. Although rare, it has been taken from off St. Lucia, Dominica, Anguilla, Guadeloupe, the Virgin Islands and Martinique.

Several specimens of *T. cubensis* in the Jeffreys collection in the National Museum of Natural History have no data with them other than their origin in the Gulf of Mexico.

Tichosina cubensis usually inhabits waters from 137 to 732 meters in the Straits of Florida, but in the Gulf of Mexico it has been taken from 3111 meters at OREGON Station 1189. The specimen from this station is a large one, illustrated on plate 27, figures 15-19. In the Caribbean the deepest record for the species is 963 meters.

Tichosina cubensis is not always easy to identify in its narrower forms or in the young. It usually maintains its great anterior width in the young but can be confused with *T. truncata, rotundovata* or *erecta,* all new species. Each of these in adult form has a character distinctly its own that readily separates it from *T. cubensis* and from each other. At a number of localities in the Caribbean only the dead shells of *T. cubensis* were found, possibly from small, defunct communities.

Localities.—G241, G242, G579, G688, G973, G974, G1036, G1102, P890, P905, P929, P930, P931, P943, P984, P991. BLAKE 167. OREGON 1189, 5927. SILVER BAY 2416, 2418, 2427. FISHHAWK 7283. West-southwest of Dry Tortugas at 213 meters (R. Cooper collection).

Types.—Figured hypotypes: USNM 109748a, 110856, 149405, 550755, 550761.

Tichosina dubia Cooper, new species

Plate 12, figure 18; plate 19, figures 2-7; plate 21, figures 1-5

About average size for the genus, rounded pentagonal in outline; maximum width anterior to midwidth; sides rounded; anterior margin broadly rounded; posterolateral extremities forming an angle of 75°-85°. Beak narrowly rounded; foramen large, moderately labiate; foramen mesothyridid to permesothyridid. Anterior commissure broadly but gently uniplicate. Lateral commissure with a dorsal curve at the anterior. Color translucent white to pale yellow. Punctae per square millimeter: 99.

Ventral valve deeper than the dorsal valve, moderately convex in lateral view and forming a low half ellipse; umbonal and median regions inflated; anterior slope flattened and forming a barely perceptible sulcus which produces a short slightly rounded tongue.

Dorsal valve less convex than the ventral valve in lateral profile; anterior profile a somewhat narrowly rounded dome with fairly long, steep sides. Median region swollen, flattening on the anterior slope and there producing a short, poorly defined fold marked by a slight depression of the flanks at the anterolateral extremities.

Ventral valve interior with small teeth and short excavated pedicle collar. Muscle scars fairly deeply impressed. Dorsal valve interior with strong socket ridges; fairly long and deeply concave outer hinge plates bordered by strongly elevated crural bases; crural processes bluntly pointed, overhanging the posterior edge of the broad, gently folded transverse ribbon; length of loop 1/3 valve length; loop flaring slightly anteriorly.

Measurements in mm	length	dorsal valve length	width	thickness	apical angle
550614a	21.7	18.8	17.7	13.4	80°
550614b	20.2	17.8	15.6	12.8	80°
64258	18.2	15.8	15.0	11.2	76°
64261 (holotype)	19.5	17.0	15.9	11.7	83°
64262	16.0	13.5	13.2	10.0	80°

Localities.—?P658. BLAKE Stations 147, 232, 254. OREGON II, Station 10513.

Diagnosis.—Medium, compact *Tichosina* with faintly uniplicate anterior commissure and large foramen.

Types.—Holotype: USNM 64261. Figured paratypes: USNM 550614a, 550603. Unfigured paratypes: USNM 64258, 64262, 550614b.

Comparison.—This species is similar to *T. erecta, rotundovata, subtriangulata* and *truncata,* all new, as its average L/W index is very similar to the average of the others. It is perhaps closest to *T. subtriangulata* with an average L/W of 1.2 (that of *T. dubia* is 1.23) but there are important differences. *Tichosina subtriangulata* is more conspicuously triangular and widens at the anterior rather than slightly narrowing as does *T. dubia.* The anterior commissure of *T. subtriangulata* is less conspicuously and consistently uniplicate than that of *T. dubia.* The foramen of *T. subtriangulata* is somewhat variable but even in the larger specimens it is smaller than that of *T. dubia. Tichosina erecta* will not be confused with *T. dubia* because that is a thin-shelled form having a L/W of 1.26, with small foramen, rectimarginate anterior commissure, fairly strongly incurved beak and finally with a very narrow loop having nearly parallel sides.

Tichosina rotundovata has a L/W index of 1.16, has more swollen dorsal valve, smaller foramen, the loop is parallel sided and not flared anteriorly like that of *T. dubia. Tichosina truncata* has a L/W of 1.19, smaller foramen, more tapering lateral extremities, rectimarginate anterior commissure and very narrow, parallel-sided loop and is thus importantly different from *T. dubia.*

Tichosina elongata Cooper, new species

Plate 12, figures 13-17; plate 33, figures 12-14

Average size for the genus, elongate and very narrowly oval in outline; greatest width near midvalve; anterior margin narrowly rounded; apical angle 68°. Beak long, narrowly rounded; foramen large, labiate, permesothyridid. Anterior commissure rectimarginate. Surface smooth, color yellowish white. Punctae numbering 75 per square millimeter.

Loop with strongly folded transverse ribbon and blunt crural processes at about mid-loop.

Measurements in mm	length	dorsal valve length	width	thickness	apical angle
550664	24.8	21.5	16.5	14.6	68°

Locality.—COMBAT Station 450. BLAKE 100 (not on map)

Diagnosis.—Elongate *Tichosina* with the width about 2/3 the length.

Types.—Holotype: USNM 550664. Figured paratype: USNM 64249.

Discussion.—The narrow form (L/W = 1.5) is unlike any other Caribbean *Tichosina*. It is suggestive of the Pacific *Liothyrella uva* (Broderip) in its elongate form, but the loop is parallel-sided rather than flaring as it is in the Pacific species.

A specimen of elongate *Tichosina* from off Morro Light, Havana, Cuba, is referred to *T. elongata* Cooper. The specimen is not a variant of *T. cubensis* because it is undeformed and almost identical to *T. elongata* except for its smaller size. The length of the specimen is 1 1/2 times the width, the same proportion as exhibited by *T. elongata*.

The specimen is not referable to *Stenosarina* because of its more elongated and protuberant beak. Unfortunately, the loop is broken but the hinge plates are like those of *Tichosina*. This species also suggests *Dallithyris sphenoidea* (Philippi), but it is even more slender than that species.

Tichosina erecta Cooper, new species

Plate 9, figures 18-27

About average medium size for the genus, elongate oval in outline with the maximum width at about midvalve; sides gently rounded; anterior margin subtruncate but with the anterolateral extremities moderately strongly rounded; posterolateral margins forming an angle of about 85°. Anterior commissure rectimarginate. Color grayish white, surface glossy and the shell thin and transparent. Shell markings consisting of concentric growth lines and faint radial lines seen only in reflected light. Beak, narrow, erect, foramen small, labiate, permesothyridid. Punctae numbering 65 per square millimeter.

Ventral valve fully twice as deep as the dorsal valve, strongly convex in lateral profile; anterior profile forming a high, narrow dome with steep sides. Umbonal and median regions strongly inflated, the anterior slope less so. Anterior near margin flattened and producing a short truncation.

Dorsal valve gently convex in lateral profile but with the maximum convexity in the posterior half; anterior profile a broad shallow dome with gently sloping sides. Umbonal region swollen, the swelling decreasing anteriorly to meet the flattening of the pedicle valve in a narrow truncated marginal area.

Ventral valve interior with a short, elevated pedicle collar and small teeth. Muscle scars moderately strongly impressed.

Dorsal valve interior with narrow loop having subparallel sides and occupying about one-third the valve length. Socket ridges short and rather thin; outer hinge plates short, moderately wide, nearly flat and margined by the elevated crural bases; outer hinge plates attached to the flattened and broad crural process which is only slightly incurved, is blunt and overhangs the posterior of the transverse ribbon. Connecting band longitudinally broad but laterally narrow and forming an angular fold medially.

Measurements in mm	length	dorsal valve length	width	thickness	apical angle
550525a (holotype)	21.5	19.0	16.6	13.3	85°
550525b	18.9	17.0	16.2	11.7	88°
550525c	19.8	17.0	15.0	12.5	81°

Locality.—G694, G938, G1125. Combat 449, 450.

Diagnosis.—Narrowly oval *Tichosina* with strongly erect beak, small foramen and short outer hinge plates in the loop.

Types.—Holotype USNM 550525a. Figured paratypes: USNM 550525b-d.

Comparison and discussion.—Compared to *Tichosina truncata* Cooper, n. sp., *T. erecta* has a deeper ventral valve, more erect beak and smaller foramen. Whereas *T. truncata* has the maximum width in the anterior, *T. erecta* is widest at midvalve and tapers strongly posteriorly; *T. truncata* is broadly truncated anteriorly and with very narrowly rounded anterolateral extremities but the same features in *T. erecta* are a very narrow anterior truncation and more broadly rounded anterolateral extremities. The exterior of the two is thus very different. The loops are also different in detail, the transverse band being narrower in *T. erecta* and with less concave outer hinge plates.

Tichosina erecta shares a similar outline with *T. bartschi* (Cooper) but its ventral valve is much deeper in lateral profile and the beak is more erect and the foramen smaller. The dorsal valve is deeper in *T. bartschi* and the profile more evenly and strongly convex.

Tichosina erecta is a much smaller shell than *T. cubensis* (Pourtalès) which has a broadly spreading anterior margin with very narrowly convex anterolateral extremities and is generally a much larger species.

Tichosina expansa Cooper, new species
Plate 6, figures 9-16

Large for the genus, roundly oval in outline with the length slightly greater than the width; sides rounded; anterior margin broadly rounded; posterolateral margins forming an angle of 90°. Lateral commissure with an abrupt anterior bend; anterior commissure broadly uniplicate. Beak low, truncated and with a fairly large foramen fairly strongly labiate and permesothyridid. Surface marked only by fine concentric lines of growth. Punctae 100 per square millimeter.

Ventral valve with unevenly convex lateral profile, the most convexity in the posterior and the anterior half forming a long flattened slope; anterior profile a broad dome somewhat narrowed medially: umbonal region narrowly swollen; flanks rounded and steep.

Dorsal valve less deep than the ventral valve and with uneven lateral profile, the posterior part strongly convex but the anterior flattened to form a long fairly steep slope; anterior profile forming a broad, somewhat flattened dome with short steep sides. Fold broad and low, originating in the anterior quarter and marked laterally by short low angulations.

Ventral valve with elevated, short pedicle collar, small teeth and visible symphytium. Muscles deeply impressed; pallial marks moderately impressed.

Dorsal valve interior with long narrow loop, extending for about one-fourth the valve length; outer hinge plates broad and flatly concave, bordered by elevated crural bases; crural processes over-hanging the broad and roundly folded transverse ribbon. Cardinal process small.

Measurements in mm	length	dorsal valve length	width	thickness	apical angle
550610	37.3	34.5	33.0	22.4	90°

Localities.—SILVER BAY 3499; P584.

Types.—Holotype: USNM 550610.

Diagnosis.—Large *Tichosina*, wide, with low wide fold and very narrow loop.

Discussion.—This species most resembles *T. bullisi* Cooper, n. sp., but differs in slightly different proportions, having a broader and less pronounced fold on the dorsal valve and in having a narrower loop. It remotely resembles *T. bartletti* (Dall) in its folding but it is a much wider shell and has a larger foramen than that species. It also suggests an unusually large *T. cubensis* (Pourtalès) but that species never attains the size of *T. expansa,* has posterolateral margins forming a smaller angle than that of *T. expansa,* has a strongly sinuate lateral commissure and a much less pronounced fold of the anterior commissure. Actually, many specimens of adult *T. cubensis* are essentially rectimarginate.

72

Tichosina floridensis Cooper, new species

Plate 11, figures 1-15; plate 12, figures 19-24

About average size (20-25 mm), elongate oval in outline, thick-shelled and with maximum width anterior of midvalve; sides well rounded; anterior mar-

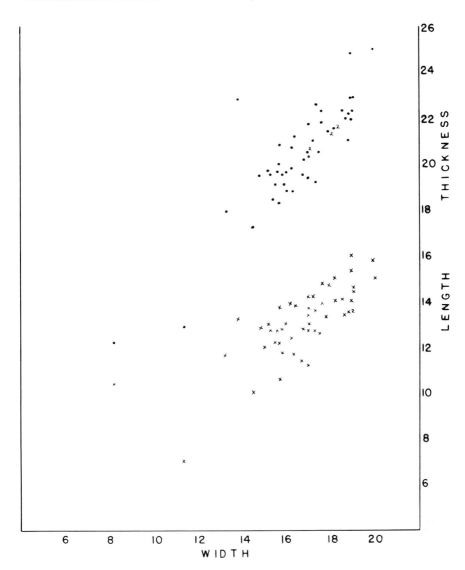

FIGURE 7. Scatter diagram showing relationship of length and thickness to width in *Tichosina floridensis,* n. sp., based on fifty specimens from Oregon Station 1025.

gin somewhat narrowly rounded; posterolateral extremities forming an angle of 72° to 101° and averaging about 88°. Valves subequally convex and nearly of equal depth. Lateral commissure straight; anterior commissure varying from rectimarginate to gently but narrowly uniplicate. Beak short, truncated, suberect; foramen small, permesothyridid, moderately labiate. Dull to glossy white; marked by concentric lines and anteriorly crowded growth varices; no radial marks seen. Punctae small, distant, numbering 89 to 100 per square millimeter.

Ventral valve slightly deeper than the dorsal valve, most convex in the posterior half; anterior profile somewhat narrowly domed and with steeply sloping sides. Umbonal and median regions swollen; anterior slope gently convex; flanks convex and steep.

Dorsal valve most convex in the posterior half and with a steep anterior slope; anterior profile narrowly domed with steeply sloping sides. Beak inserted under ventral symphytium; posterior third strongly swollen. Folding of dorsal valve indistinct, visible at anterior of some specimens as a slight nasuteness.

Ventral valve interior with fairly large, elongated teeth supported by shell thickening under them; pedicle collar very short; muscles deeply impressed; pallial trunks strongly marked and visible through the translucent shell.

Dorsal valve with loop occupying 1/3 the valve length, nearly parallel-sided, narrow; outer hinge plates wide and deep; crura short, forming a high ridge along the inner edge of the outer hinge plates; crural processes bluntly pointed, posterior to transverse band which is only moderately broad and moderately strongly folded medially.

Measurements in mm	length	dorsal valve length	maximum width	thickness	apical angle
550738a (holotype)	23.8	21.4	20.5	14.9	93°
550737-23	25.0	22.3	19.9	15.8	91°
550737-33	24.8	21.4	18.9	16.0	72°
550737-11	22.9	20.9	19.0	14.6	93°
550737-3	24.1	18.8	17.9	13.3	90°
550737-15	21.8	19.6	18.9	14.0	101°
550737-17	20.5	18.2	17.0	13.4	89°
550737-47	19.5	17.6	16.7	11.4	95°
550737-45	18.3	16.8	15.7	10.6	90°
550737-28	17.2	15.6	14.4	10.0	90°
550737-37	12.9	11.7	11.4	7.0	96°
64248a	19.7	17.8	17.5	12.5	96°
64248b	19.4	17.3	16.4	12.3	91°
64248c	18.7	16.9	15.9	11.2	96°

Localities.—BLAKE Station 45; OREGON Stations 955 and 1025. Southwest of Egmont Key, mouth of Tampa Bay, Florida. Off Havana, Cuba at 119 fathoms (=218 meters). West-southwest of Dry Tortugas at 183 meters (R. Cooper collection).

Diagnosis. — Medium size, elongate oval, small foramen, subequally convex valves, thick shell and often vitreous luster.

Types.—Holotype: USNM 550738a. Figured paratypes: USNM 64248a, b, 550737e, 550737-(15). Unfigured paratypes: USNM 64248c, 550737a-d, f-h, 550738b. Measured paratypes: USNM 550737-(1-14)(16-51).

Comparison.—As this species has the form, outline and size of an average *Tichosina* it must be compared with a fair number of species: *T. bartschi* Cooper, *dubia*, n. sp., *erecta*, n. sp., *labiata*, n. sp., *obesa*, n. sp., *plicata*, n. sp., *rotundovata*, n. sp., *subtriangulata*, n. sp., and *truncata*, n. sp. It differs from *T. bartschi* in its more rounded, solid form, lustrous shell, less tapered beak and tendency for the adult to be uniplicate.

Tichosina dubia is a slightly smaller species than *T. floridensis,* with less convex valves, larger foramen and strong anterior truncation. *Tichosina erecta* is an entirely differently shaped shell than *T. floridensis* and suggests a narrow *T. cubensis*(Pourtalès). It is thus fairly narrow, strongly inequivalve, with a broadly rounded anterior, rectimarginate anterior commissure with no tendency to uniplication. *Tichosina labiata* is quite unlike *T. floridensis* because it is triangularly oval in outline, has a large foramen and wide, fairly strongly uniplicate, anterior commissure.

Tichosina obesa is very similar in appearance to *T. floridensis* but attains a maximum size of only 20 mm, has subequally convex valves but a stronger tendency to uniplication than *T. floridensis.* The foramen of *T. obesa,* although small, is generally larger than that of *T. floridensis* and the beak is more incurved. The size difference and tendency to fold at a fairly early stage will help to separate the two in large collections but separation of individuals may be difficult.

Tichosina plicata is readily separable from *T. floridensis* because it has a large foramen and very strong anterior folding. *Tichosina rotundovata* is about the same size as *T. floridensis* but has a thin and delicate shell with a tendency to a triangular outline. It is broadly rounded at the anterior margin, with no tendency toward nasuteness as in *T. floridensis.* It has a larger foramen and is completely rectimarginate. The broad anterior, broad uniplication and large foramen of *T. subtriangulata* will readily separate that species from *T. floridensis. Tichosina truncata* is another subtriangular species having an entirely different shape from that of *T. floridensis.*

Tichosina floridensis is best developed in the Gulf of Mexico. Specimens taken by the BLAKE off Havana, Cuba, clearly belong here.

Tichosina labiata Cooper, new species

Plate 14, figures 11-20

About average size for the genus, slightly wider than long and widest just anterior to midvalve; sides strongly rounded; anterior margin broadly rounded; posterior margins forming an angle of 72°-78°. Anterior commissure broadly but slightly uniplicate; lateral commissure curved anteriorly.

Beak fairly long; foramen large and strongly labiate, permesothyridid. Punctae numbering 102 per square millimeter.

Ventral valve broadly and moderately convex in lateral and anterior profiles; lateral slopes rounded and steep. Dorsal valve with broadly domed anterior profile, somewhat less convex than that of the ventral valve.

Dorsal valve with loop equal in length to about one-third the valve length; loop expanding slightly anteriorly and with a very broad, slightly and broadly folded transverse band. Outer hinge plates large, wide, and deep and margined by strongly elevated crural bases.

Measurements in mm	length	dorsal valve length	width	thickness	apical angle
550577a (holotype)	25.3	21.9	21.7	15.8	78°
550577b	23.3	20.7	19.9	14.0	72°

Locality.—P876.

Diagnosis.—Triangularly oval *Tichosina* with large foramen.

Types.—Holotype: USNM 550577a; unfigured paratypes: USNM 550577b.

Discussion.—This species is smaller than *T. cubensis* (Pourtalès) and is definitely uniplicate and never so strongly triangular as *T. cubensis*. It differs from *T. subtriangulata*, n. sp., in being more convex, more oval than triangular, more anteriorly rounded and with much rounder lateral margins as well as a larger foramen.

This species suggests *T. truncata* and *T. rotundovata*, n. spp., but differs from both. From the first it differs in having a larger foramen, more oval outline and its maximum width near midvalve. From *T. rotundovata* it differs in having a wider loop with a stronger transverse band, a larger foramen and less swollen dorsal valve. It differs from *T. bartschi* Cooper in its widely rounded anterior.

Tichosina martinicensis (Dall)

Plate 8, figures 1-9

Gryphus moseleyi Dall, 1920: 318-319 = *Gryphus martinicensis* Dall, (specimen 64255 only). Not *Terebratula moseleyi* Davidson, 1878.

Almost circular in outline but with the pedicle valve slightly longer than the dorsal; about average size; sides and anterior margin broadly rounded apical angle 88°. Anterior commissure rectimarginate. Beak low, narrowly rounded, erect and strongly truncated; foramen small, labiate; symphytium nearly completely hidden by the ventral beak. Surface marked by concentric lines of growth and obscure, interrupted radial lines on the flanks. Body cavity small and confined to the posterior third of the shell. Punctae numbering 158 per square millimeter in the type. Ventral valve deeper than the dorsal valve, fairly strongly convex in lateral profile with the maximum convexity in the posterior half, anterior profile broadly and gently convex; beak and median region swollen; lateral slopes convex and gentle.

Ventral valve interior with fairly large teeth, and with four well developed pallial trunks branching near the anterior margin. Dorsal valve with a short, narrow loop whose length is one-quarter that of the valve length; socket ridges stout; outer hinge plates narrow, short without elevated margins; crural processes short, stout, bluntly pointed. Loop anterior to crural processes short; transverse ribbon broad and medially folded.

Measurements in mm	length	dorsal valve length	width	thickness	apical angle
Holotype	23.2	20.6	21.0	13.9	88°

Locality.—BLAKE Station 193.

Diagnosis.—Nearly circular *Tichosina* of average size.

Types.—Holotype: USNM 64255.

Discussion.—This species is characterized by its nearly circular form and in this respect is very similar to *Liothyrella moseleyi* (Davidson). In fact, Dall (1920: 319) sent the type specimen to Davidson, who confirmed this identification. Nevertheless, Dall confessed some doubt about the identification of his West Indian specimen with one from Kerguelen Island on the other side of the world. He therefore suggested the name *"martinicensis"* but never formally described the species. Dall's specimen differs from *L. moseleyi,* according to figures given by Davidson (1886: plate 2, figures 1-4), in having a more extended ventral beak, larger foramen, a more attenuated dorsal beak and a wider loop with narrower transverse ribbon. The loop as figured by Davidson is like that of *Liothyrella* rather than that of *martinicensis.* The measurements of *L. moseleyi* are almost identical to those of Dall's species, but the beak regions of the two are quite unlike.

Dall mentions two other lots of specimens as belonging to *"Gryphus moseleyi."* One of these is a smaller but more nearly circular specimen (USNM 110853) from off the New Jersey coast. This is a deep water form as it comes from 1362 fathoms (=2492 meters). The loop of this specimen is broken but it is short and with short inner hinge plates. The punctae count in one square millimeter is 84, much less than that of the type. This specimen (USNM 110853) suggests a young example of *T. cubensis*(Pourtalès)but the great depth at which it was taken raises some doubt. It is referred to *Abyssothyris* with a query because the anterior commissure is slightly sulcate.

The second lot placed in his species by Dall is USNM 110887, cited in his publication (1920: 318) as from the Gulf of Panama but actually from ALBATROSS Station 3370, near Cocos Island. The even single valves are unlike any *Liothyrella* or other short-looped terebratulid from that part of the Pacific. The best valve in the lot is oval and measures 38 millimeters long and 31 millimeters wide. It is thick-shelled and has a large labiate foramen suggestive of the type specimen of *Liothyrella uva* from the Gulf of Tehuantepec to the northwest. The loop is unknown and the cardinalia not well preserved. These specimens certainly have nothing to do with *L. moseleyi* or *T. martinicensis.*

77

Tichosina obesa Cooper, new species
Plate 9, figures 1-17

Small for the genus, elongate oval in outline with the greatest width at mid-valve; valves unequally deep, the ventral valve having the greater depth. Sides rounded; anterior margin narrowed and often nasute. Posterolateral extremities forming an angle between 70° and 90°. Anterior commissure variable, ranging from rectimarginate to strongly uniplicate, the latter condition occurring in the majority of adult specimens. Umbonal region narrowly extended beyond the hinge 1 to 3 mm. Beak small and narrow, usually erect; foramen small, labiate, permesothyridid in position. Symphytium mostly obscured. Surface smooth; white in color; growth lines fairly strong, anteriorly crowded. Punctae numbering 135 per square millimeter in a young specimen (550516p).

Ventral valve evenly and strongly convex in lateral profile and narrowly domed in anterior profile; umbonal region narrowly swollen, the swelling extending to midvalve; anterior half somewhat flattened and with a steep anterior slope; tongue short and rounded.

Dorsal valve less convex in lateral profile than the ventral valve and with the maximum convexity in the umbonal region. Anterior profile strongly domed and with very steep slopes. Median region inflated, the inflated part extending anteriorly and somewhat flattened to form an inconspicuous fold that is defined mostly at the margin.

Ventral interior with a short, elevated pedicle collar and small, elongated teeth. Muscle region strongly impressed.

Dorsal valve interior with stout cardinalia; socket ridges thin and elevated; cardinal process fairly large, semicircular in outline; outer hinge plates deeply concave and bounded by strongly elevated crural bases; crura short and flat; crural processes bluntly pointed and overhanging the transverse band which is longitudinally broad, with a narrow, gentle median fold and small posterior notch. Loop occupying one-third the valve length from the beak.

Measurements in mm	length	dorsal valve length	width	thickness	apical angle
550514a	18.0	15.5	13.5	12.6	81°
550514b	16.6	14.0	13.0	11.0	81°
550514c	15.4	13.6	13.0	10.0	85°
550585a (holotype)	16.2	13.9	12.5	11.5	79°
550585b	16.2	13.9	11.7	11.2	71°
550585c	13.7	12.3	12.7	8.8	88°
550585d	15.4	13.3	13.0	10.6	81°
550516a	14.2	12.3	11.5	9.5	72°
550516b	14.6	12.8	11.1	9.4	72°

Localities.—P479, P584, P650, P694, P707, P708, P734, P736, P737, P739; OREGON 2232, 4297, 4459, 4461, 4466.

Diagnosis.—Small, oval, thick *Tichosina* with stout, thick loop.

Types.—Holotype: USNM 550585a; figured paratypes: USNM 550514a, b, 550585b, d, e. Measured paratypes: USNM 550514a-c, 550585a-d, 550516a, b.

Comparison and discussion.—This species is distinctive for its relatively small size, strong convexity, fairly strong adult uniplication and small foramen. It does not compare with any of the larger shells but does have similarity to *T. bahamiensis* and *T. abrupta,* n. spp. It differs from the first in having a different shell texture, that of *T. obesa* being somewhat chalky in old shells but seldom translucent as in the Bahama species. The latter is rectimarginate, much smaller and with less rounded sides than *T. obesa. Tichosina abrupta* is more nearly the average size of *T. obesa* but it is only faintly uniplicate, its beak lower; the loop is anteriorly narrower than that of *T. obesa* and the outer hinge plates are much longer and deeper.

Tichosina ovata Cooper, new species

Plate 30, figures 1-20

Large, thin-shelled, elongate oval in outline, the width about 5/6 the length and greatest slightly anterior of midvalve; sides rounded; anterior margin broadly rounded. Posterolateral extremities forming an angle of 70° to 95°. Anterior commissure varying from moderately to strongly uniplicate. Beak suberect, short, truncated; foramen moderately large, with thick rim and slightly labiate, permesothyridid. Surface marked by concentric lines of growth and faint radial lines on the sides. Color white underneath a thin, rich brown periostracum. Punctae numbering 78-100 per square millimeter.

Ventral valve strongly and evenly convex in lateral profile, with the maximum convexity at about midvalve. Anterior profile roundly convex and with steeply sloping sides. Umbonal region swollen, the swelling merging quickly into the general valve convexity which is fairly strong. Sulcus prominent only at the anterior where the front margin is extended as a short gently rounded to nearly straight tongue. Sulcus wide, occupying not quite half the valve width and defined by a short low plication on each side.

Dorsal valve less convex and deep than the ventral valve, gently convex in lateral and anterior profiles. Umbonal region moderately swollen, the swelling continuing into the indistinct fold that can be seen starting from a half to three-quarters the length from the beak. Anterior folded portion flattened and elevated slightly above the rounded and convex flanks; fold defined by a low wrinkle on each side, strong in some specimens, weak in others, especially the young.

Ventral valve interior with short anteriorly excavated pedicle collar, small elongated teeth; muscles lightly impressed.

Dorsal valve with small, wide cardinal process; loop nearly parallel-sided, stout, equal in length to 1/4 to 1/3 the valve length. Socket ridges thin; outer hinge plates short, deeply concave and margined by the strongly elevated crural bases: crura short; crural processes bluntly pointed, over-hanging the transverse ribbon, the latter broad and with a strong median fold; anterolateral extremities not extended with small points.

Measurements in mm	length	dorsal valve length	maximum width	thickness	apical angle
549434a (holotype)	36.4	32.6	30.1	21.6	87°
549434b	35.4	31.5	28.3	20.0	90°
549434c	35.0	30.0	25.9	20.0	81°
549434d	32.5	28.8	26.4	18.7	84°
549434e	33.0	29.5	27.5	18.6	95°
549434f	37.0	33.7	34.0	21.7	93°
549434g	37.7	32.0	28.5	22.8	82°
549434h	31.1	27.8	26.8	17.0	89°
549434i	35.9	31.4	26.0	20.3	83°
549434j	31.8	28.5	26.0	19.6	93°
549434k	35.0	30.7	28.8	19.4	92°

Locality.—OREGON Station 1408.

Diagnosis.—Large, uniplicate *Tichosina* with moderately strong fold and slightly labiate foramen.

Types.—Holotype: USNM 549434a. Figured paratypes: USNM 549434b-d, 1-n. Unfigured paratypes: USNM 549434e-k.

Comparison and discussion.—This is a large species that needs comparison only to the larger members of the genus: *Tichosina cubensis* (Pourtalès), *T. ? bartletti* (Dall), *T. bullisi* Cooper and *T. expansa* Cooper. It differs from *T. cubensis* in its strongly oval outline, stronger uniplication, nearly straight lateral commissure, less convex dorsal valve and smaller foramen. Large specimens of *T. cubensis* are rather rotund and their rounded triangular outline is very distinctive and very different from the elongate oval outline of *T. ovata. Tichosina ? bartletti* is about the same size but is more strongly folded and tapers posteriorly more strongly than *T. ovata.* The foramen is more labiate than that of *T. ovata* and the ventral valve has a much longer and more conspicuous anterior tongue. *Tichosina ovata* differs from *T. bullisi* in outline, the latter having much more rounded sides, lesser development of the fold and less narrowed posterior. The maximum width of *T. bullisi* is at about midvalve whereas that of *T. ovata* is anterior to midvalve. *Tichosina expansa* is widely rounded with the maximum width at about midvalve and a broadly rounded anterior and is thus quite unlike *T. ovata.*

The count of punctae per square millimeter of *T. ovata* is variable. Two counts made on different parts of the umbonal slope of a specimen (USNM 549434h) somewhat younger than the majority yielded the numbers 78 and 83. Two adult specimens on which counts were made (USNM 549434e, k) proved to have higher counts: 94 and 99 on the former and 100 on the latter. This suggests the possibility of the punctae becoming denser with age or varying between specimens.

Tichosina ovata is confined to the Gulf of Mexico.

Tichosina pillsburyae Cooper, new species

Plate 26, figures 12-27

Large for the genus, longer than wide, oval in outline; greatest width just anterior to midline. Color white to translucent. Sides rounded; posterolateral margins forming an angle of 73° to 90°. Beak suberect to erect; foramen small, strongly labiate, permesothyridid; anterior commissure strongly uniplicate. Valves subequally deep, both fairly deep, but the dorsal valve less deep than the ventral one. Punctae numbering 76-95 per square millimeter.

Ventral valve strongly convex in lateral profile, the maximum convexity in the posterior part, the anterior slope convex and steep. Anterior profile forming a broad evenly rounded dome having steep slopes. Umbonal region narrowed, swollen the swelling extending to midvalve. Anterior forming a short, truncated tongue very slightly depressed below the flanks.

Dorsal valve gently and evenly convex, much less convex than the opposite valve in lateral view; anterior profile forming a broad dome somewhat narrowly convex medially but with long, steep slopes. Median region strongly swollen; anterior slope flattened and forming a conspicuous fold occupying about 3/5 the valve width. Flanks convex and depressed below the surface of the fold.

Ventral valve much thickened internally; pedicle collar short; teeth small but buttressed by strong thickenings; diductor patch oval, deeply impressed; pallial trunks narrow, deep.

Dorsal valve having small elliptical cardinal process, short socket ridges and deeply concave, short outer hinge plates. Crural bases forming a wall along the inner edge of the outer hinge plates; crura moderately long; crural processes blunt and overhanging the transverse ribbon of the loop; transverse ribbon stout and with a moderate median fold. Adductor scars elongate not extending anterior to the anterior end of the loop. Length of loop about 1/3 valve length.

Measurements in mm	length	dorsal valve length	width	thickness	apical angle
550622a	30.9	27.2	25.2	20.0	85°
550622b (holotype)	26.7	22.6	20.8	18.7	78°
550622c	27.0	23.7	19.7	17.3	73°

Locality.—Station P1387, P1393.

Types.—Holotype: USNM 550622b. Figured paratypes: USNM 550622a, c.

Diagnosis.—Fairly large *Tichosina* having a small, strongly labiate foramen and strongly uniplicate anterior commissure.

Comparison.—Of all the species known from the Caribbean region this one most strongly resembles two strongly uniplicate forms: *T. ? bartletti* (Dall) and *T. plicata,* new species. It differs from the former in its much smaller size, more erect beak, less swollen dorsal valve, in the greater thickening of the interior by adventitious shell, more elongated outer hinge plates, and in having a stronger fold when compared to specimens of comparable size. *Tichosina plicata* is smaller than *T. pillsburyae,* has a much larger foramen, earlier development of the fold, entirely differently shaped loop, and a more convex dorsal valve. Named for R/V JOHN ELLIOTT PILLSBURY.

Tichosina plicata Cooper, new species

Plate 5, figures 9-15; plate 21, figures 6-11

Average size for the genus, oval in outline; inequivalve, the ventral valve having the greater depth; sides anteriorly rounded and anterior margin narrowly rounded; posterolateral margins tapering and forming an angle of about 75°-85°. Lateral commissure curved anteriorly; anterior commissure strongly uniplicate. Beak short, suberect, moderately truncated; foramen fairly large, labiate, mesothyridid to permesothyridid. Surface marked by strong concentric growth lines and faint interrupted radial lines on the flanks. Punctae numbering 132 per square millimeter.

Ventral valve strongly convex in lateral profile and most convex posterior to midvalve, flattened anteriorly, anterior profile a low broad dome with short, gently sloping sides; umbonal and median regions swollen; anterior somewhat flattened; anterior tongue fairly long and narrowly rounded.

Dorsal valve fairly deep, strongly and evenly convex with the maximum convexity near midvalve; anterior profile a strongly convex, narrow dome with precipitous sides. Umbonal and median regions swollen, the swelling continuing anteriorly to form a low, narrow fold Flanks steep.

Pedicle valve with small teeth and short pedicle collar anteriorly excavated. Dorsal valve with small, transverse cardinal process; socket ridges stout; outer hinge plates fairly long and wide, with concave surface and margined by the crural base which forms an elevated rim on the inner edge of the hinge plate; crural processes short, flat and fairly sharply pointed and overhanging the transverse ribbon. Anterior of the loop flaring; transverse ribbon broad and broadly but moderately folded medially.

Measurements in mm	length	dorsal valve length	width	thickness	apical angle
550607a (holotype)	23.5	20.1	18.7	15.6	76°
550607b	25.5	22.1	19.8	17.7	80°
550607c	22.4	19.5	18.2	15.3	83°
550607d	23.2	19.8	18.0	14.6	75°

Localities.—OREGON 5624; P838?

Types.—Holotype USNM 550607a; measured paratypes USNM 550607b-d;

82

figured paratype USNM 550607e. Figured specimen: USNM 550588.

Diagnosis.—Strongly folded *Tichosina* of average size.

Discussion.—Although this species is barely half the size of *T.* ? *bartletti* (Dall) it has been mistaken for that species. Beside size, there are many other differences from *T.* ? *bartletti*; the dorsal valve is more convex in lateral profile and is more narrowly swollen in anterior profile; the ventral tongue is longer, the apical angle is larger and the loop has more divergent descending elements. Young *T.* ? *bartletti* comparable in size to *T. plicata* are much more rounded, flatter and have a rectimarginate anterior commissure, a shell entirely different from the compact plicated species from Venezuela.

The loop of a specimen from P838 is identical to that of the specimens from OREGON 5624. The PILLSBURY specimen is not so strongly folded but it is probably somewhat younger than those collected by the R-V ORE-GON.

Tichosina rotundovata Cooper, new species
Plate 7, figures 18-32

Roundly but elongate oval in outline with the maximum width just anterior to midvalve; sides rounded and anterior margin broadly rounded; posterolateral margins forming an angle near 80°. Shell white, thin, translucent, glossy, with concentric and very faint radial markings, the latter mostly on the flanks. Beak low, narrow, suberect; foramen small, labiate, permesothyridid. Anterior commissure rectimarginate. Punctae numbering 82 to 100 per square millimeter.

Ventral valve deeper than the dorsal valve, moderately convex in lateral profile with the maximum convexity in the umbonal region; anterior profile broadly and moderately domed with moderately steep slopes. Median region narrowly swollen, the swelling extending to mid-valve anterior to which it declines.

Ventral valve moderately and evenly convex in lateral profile and broadly and moderately domed in anterior profile. Median region swollen, but the swelling declining on the anterior slope which is gently convex; lateral slopes short but steep.

Ventral valve with short, elevated pedicle collar and lightly impressed muscle scars.

Dorsal valve interior with short loop, occupying a quarter of the valve length; sides subparallel. Socket ridges short and slender; outer hinge plates large, long and nearly flat; crural bases strongly elevated and forming a partition along the inside edge of the outer hinge plate; crural processes broad, bluntly pointed and overhanging the posterior edge of the transverse ribbon, which is moderately broad longitudinally and folded medially.

Measurements in mm	length	dorsal valve length	width	thickness	apical angle
550526a	21.0	19.3	18.0	13.8	81°

550526b	21.1	19.0	18.3	12.5	81°
550526c	19.9	18.0	17.0	12.0	85°
550526d	20.8	18.9	18.0	12.8	80°
550526e	21.8	19.6	17.7	13.3	84°
550526f (holotype)	21.9	19.8	18.0	14.0	84°

Localities.—G252, G482, G503, G942, G1312, G1314.

Diagnosis.—*Tichosina* having the appearance of a small *T. cubensis* (Pourtalès) but narrower and with a short loop.

Types.—Holotype: USNM 550526f; figured paratypes: USNM 550526a, g, h. Unfigured paratypes: USNM 550526b-e.

Comparison and discussion.—As mentioned in the diagnosis this species suggests *T. cubensis* but is smaller and much narrower. Young *T. cubensis* retain fairly faithfully the broad outline of the dorsal valve but they are much broader than specimens of *T. rotundovata* of the same size.

This species has a different shape, color, profile and outline when compared with *T. bartschi* (Cooper). It differs from *T. erecta,* n. sp., in its wider shell, more convex dorsal valve, larger foramen and less erect beak. *Tichosina truncata,* n. sp., suggests *T. rotundovata* in size, color and general appearance but it differs in shape and loop, having its greatest width well anterior to the middle, in having a slightly convex to truncated anterior with narrowly rounded anterolateral extremities and a longer loop.

Tichosina solida Cooper, new species
Plate 9, figures 28-34, plate 25, figures 23-30

Fairly large for the genus, longer than wide and elongate, oval in outline, with maximum width at the middle; lateral commissure straight except for a small anterior angulation; anterior commissure broadly and gently uniplicate. Beak narrow, not strongly protruding; foramen large, moderately labiate, permesothyridid; surface marked by fairly strong concentric growth lines. Punctae number 78-84 per square millimeter.

Ventral valve deeper than the dorsal valve, moderately convex in lateral profile with the anterior flattened; anterior profile a broadly convex dome with steep sides; umbonal and median regions swollen; anterior slope short and steep; tongue short and broadly truncated.

Brachial valve evenly and moderately convex, with the maximum convexity at about midvalve; anterior profile a broad dome similar to that of the ventral valve but not deep; median region swollen moderately; anterior slope somewhat flattened.

Ventral valve interior with short, excavated pedicle collar; muscle marks not strongly impressed. Dorsal valve having a loop equal in length to about 1/3 the valve length; loop narrow, flaring slightly; outer hinge plates wide and concave, margined by elevated crural bases; crura long and flat, crural processes sharp, overhanging dorsal edge of the longitudinally broad transverse ribbon; fold of transverse ribbon, sharp and narrow. Cardinal process fairly large.

84

Measurements in mm	length	dorsal valve length	width	thickness	apical angle
549433a (holotype)	30.3	26.4	23.2	19.9	81°
549433b	26.6	23.0	20.5	17.1	85°
549433c	21.6	19.0	17.8	12.3	77°

Localities.—G1029; EOLIS Station 1; SILVER BAY 2418, 2426.

Types.—Holotype: USNM 549433a. Figured paratype: USNM 550619, 549433b. Unfigured paratype: USNM 549433c.

Diagnosis.—Elongate oval *Tichosina* with slightly uniplicate anterior commissure.

Discussion.—The specimens described under this heading strongly suggest *T. cubensis* (Pourtalès) in the large foramen and thick shell, but they have an entirely different shape. *Tichosina cubensis* is a strongly triangular shell with the maximum width well anterior of the middle. Furthermore its adult lateral commissure is strongly sinuate and the anterior commissure is usually rectimarginate or nearly so, seldom folded as much as that of the Sand Key specimens.

A specimen (USNM 550619a) from locality G1029 has an aberrant loop suggesting repair or superposed growth. The outer hinge plates are very deep and the marginal rims formed by the crural bases are exceptionally high. The main part of the loop appears completely normal and characteristic for the species, but at the anterior end of each hinge plate a narrowly bent ribbon of calcite is given off. This is narrowly curved and its opposite end attached to the descending branch of the loop at the posterior end of the transverse ribbon. The crural process is marked by a thickening or ridge on its outside that extends to the point. This ridge extends proximally and joins the hinge plate and forms its anterior margin. The anterior margin of the hinge plate is slightly deflected toward the ventral valve. The structure suggests that the original loop has been overgrown by a second more extended one which was laid down inside the old loop and extended anterior to it. The bent ribbons at the anterior end of the hinge plate may represent the remnants of the transverse ribbon now overlaid by the new loop (see plate 25).

Tichosina subtriangulata Cooper, new species

Plate 7, figures 1-17

About average size for the genus, subtriangular in outline; maximum width anterior to midvalve in the majority of specimens; sides gently rounded; anterolateral extremities narrowly rounded; anterior margin faintly convex to nearly straight. Apical angle between 70° and 80°. Beak short, suberect, strongly truncated; foramen small, labiate, permesothyridid. Lateral commissure nearly straight to slightly curved; anterior commissure varying from rectimarginate to broadly uniplicate. Surface marked by fairly strong concentric lines of growth and obscure but often numerous, interrupted, very fine, elevated radial lines concentrated on the flanks. Color yellowish, with a high gloss where the thin epidermis is worn. 102 punctae per square millimeter.

Ventral valve deeper than the dorsal valve, moderately convex in lateral profile with the maximum convexity near midvalve; anterior profile broadly and moderately convex with gently sloping sides. Anterior slope convex and steep.

Dorsal valve with gentle lateral profile but with broadly convex anterior profile about equal to that of the ventral valve. Beak low and rounded, concealing the symphytium. Flanks convex.

Ventral interior with long slender teeth; pedicle collar short, anteriorly excavated. Pallial trunks four, narrow and widely divergent.

Dorsal valve having a long, narrow loop extending one-quarter the valve length into the shell; outer hinge plates narrow and deeply concave; crural bases forming a high ridge bounding the outer hinge plates; crural processes located at end of hinge plates; transverse ribbon very broad and with a low, narrow median fold.

Measurements in mm	length	dorsal valve length	width	thickness	apical angle
226290-5	22.8	20.1	18.5	12.8	72°
-41 (holotype)	22.4	19.9	18.7	13.4	78°
-19	21.4	18.8	18.0	13.0	75°
-10	20.2	17.7	17.3	12.4	77°
-21	19.0	16.5	15.6	11.3	74°
-7	18.3	15.6	14.8	11.4	69°
-28	15.9	14.1	13.2	8.1	75°
-35	11.8	10.6	10.0	6.1	84°
-34	9.0	8.0	7.9	4.3	90°

Localities — ALBATROSS Stations 2152, 2343; FISH HAWK Station 6070.

Types.—Holotype: USNM 226290-41; figured paratypes 226290-5, 226290-10, 226290-23, 226290a, e.

Diagnosis.—Medium sized *Tichosina* with maximum width anterior to the middle, anterior commissure usually rectimarginate and shell subtriangular in outline.

Discussion.—This species resembles a small *T. cubensis* (Pourtalès) and is undoubtedly related to it. It is a much smaller and less convex species usually without the anterior broad faint uniplication of *T. cubensis*. It is also yellower and has a fairly lustrous shell. Other differences are to be seen in the shape which is subtriangular rather than subpentagonal to triangular as *T. cubensis*.

Perhaps the closest species to *T. subtriangulata* is *T. bartschi* Cooper which is about the same size and has a similar lustrous and yellowish exterior but whose shape is quite different. *Tichosina bartschi* has a tapering anterior and the maximum width is at about midvalve whereas the maximum width of *T. subtriangulata* is anterior to midvalve and the shell is distinctly more triangular. Internally the loops of the two are very similar.

86

Tichosina truncata Cooper, new species
Plate 3, figures 6-9, plate 12, figures 1, 2

About average size, elongate oval to elongate triangular in outline; side gently rounded; anterolateral extremities narrowly rounded; anterior margin truncated; posterolateral margins gently concave and forming an angle of 70° to 90°. Greatest width in the anterior half. Valves unequally deep, the ventral valve having the greater depth; anterior commissure rectimarginate. Beak narrow; foramen small, labiate, permesothyridid. Color white, translucent and dull glossy, marked only by concentric lines of growth. Punctae per square millimeter 63-72.

Ventral valve moderately convex in lateral profile; broadly domed in anterior profile with the median region gently convex and the lateral slopes short and steep. Umbonal and median regions swollen, the swelling decreasing anteriorly; anterior slope flattened; flanks narrow and steep.

Dorsal valve gently convex in lateral profile, less convex than the ventral valve; anterior profile a broad dome somewhat more convex than the ventral valve in the same profile; sides short and steep. Umbonal and median regions swollen; anterior slope flattened but not so strongly as the anterior slope of the other valve.

Pedicle valve with small elongated teeth, a short pedicle collar and moderately deeply impressed muscle scars. Dorsal valve with thin, erect socket ridges and narrow sockets; outer hinge plates broad and moderately concave, attached to the crural base which is flat and forms an elevated margin to the inner edge of the outer hinge plate; crural processes curved inward, bluntly pointed and overhanging the transverse ribbon which is longitudinally broad and narrowly folded medially; loop narrow and with nearly parallel sides, occupying one-third the valve length.

Measurements in mm	length	dorsal valve length	width	thickness	apical angle
550524a (holotype)	20.2	20.0	17.5	13.7	78°
550524b	19.5	17.6	16.8	11.3	83°
550524c	16.1	14.8	13.3	8.9	75°
550524d	22.4	20.3	18.0	13.2	75°

Localities.—G708, G898, G947; P584, P587, P594.

Diagnosis.—Subtriangular to elongate oval *Tichosina* with truncated anterior margin and rectimarginate anterior commissure.

Types.—Holotype: USNM 550524a. Figured paratype: USNM 550587. Unfigured paratypes: USNM 550524b-d.

Comparison and discussion.—This species at once suggests the young of *Tichosina cubensis* (Pourtales) but these are usually rounder and wider than *T. truncata*. The exterior outline and profile of *T. cubensis* are very distinctive. The anterior margin is usually broadly rounded rather than truncated and the sides are broadly rounded.

This species also suggests similarity to *T. bartschi* (Cooper) in its outline and profile, but the latter species is rather more slender in outline and has the maximum width at about midvalve. *Tichosina truncata* has a deeper ventral valve and shallower dorsal valve than *T. bartschi*, the type of which is almost exactly the same length as adults of *T. truncata*. *Tichosina bartschi* differs in color from *T. truncata*, as it is yellowish and glossy rather than white.

Tichosina sp. 1
Plate 12, figures 8-12

About medium in size, roundly oval in outline with the valves sub-equally convex, the ventral valve slightly more convex than the other. Maximum width anterior to midvalve; apical angle 85°. Beak narrowly rounded, slightly extended; foramen small, strongly labiate, submesothyridid. Anterior commissure with a narrow dorsal wave; lateral commissure straight. Color translucent white. Punctae per square millimeter: 89.

Ventral valve strongly convex in both profiles but most convex in anterior profile; median region inflated; anterior slope with a narrow, poorly defined sulcus in the anterior third.

Dorsal valve strongly convex in both profiles; median region inflated; anterior slope steep, convex and folded at the anterior to accommodate the slight tongue of the ventral valve. Loop narrow, transverse ribbon with a strong median fold.

Measurements in mm	length	dorsal valve length	width	thickness	apical angle
550665	24.9	22.1	21.4	16.9	85°

Locality.—OREGON Station 6715.

Diagnosis.—Rounded and inflated *Tichosina* with narrowly folded anterior commissure.

Types.—Figured specimen: USNM 550665.

Discussion.—This species is represented by a single specimen, which is unlike any other in the collection in its inflated valves and narrowly folded anterior commissure. It partakes of the form of *T. cubensis* (Pourtalès) but is rather more roundly pentagonal than triangular and the anterior fold is unlike any aberration of the anterior commissure seen in *T. cubensis*.

Tichosina sp. 2

This is an elongate oval, large specimen 29 mm long, 24 mm wide and 18 millimeters thick and suggesting *T. cubensis* (Pourtalès). It differs from that species in not being flared anteriorly and having a gently rounded anterior margin rather than a truncated one. Furthermore the dorsal valve is deeper and the umbonal region less swollen than that of *T. cubensis*. The commissure is folded to about the same degree as that of *T. cubensis*.

Locality.—OREGON Station 6715.

Diagnosis.—Large *Tichosina* with swollen dorsal valve and sides narrowing anteriorly.

Types.—Described specimen: USNM 550666.

Tichosina sp. 3

Plate 18, figures 8-11

Small, narrowly oval in outline with the valve subequal in depth; greatest width anterior to midvalve; lateral commissure anteriorly curved; anterior commissure strongly narrowly uniplicate. Beak short; foramen large, labiate and permesothyridid. Color yellowish white. Loop with moderately long and deeply concave outer hinge plates. Transverse ribbon broken.

Measurements in mm	length	dorsal valve length	width	thickness	apical angle
64260	17.5	15.0	13.5	12.0	75°

Locality.—BLAKE Station 253.

Types.—Figured specimen: USNM 64260.

Discussion.—This species suggests *T. obesa* Cooper, n. sp., from off Venezuela but the two specimens from Grenada do not compare in detail with *T. obesa*. The foramen of *T.* sp. 3 is larger than the majority of specimens of *T. obesa*; the contours are different, the maximum width of the Grenada specimens is anterior whereas that of the Venezuela species is median. Although the profiles are similar, the anterior folding of the Grenada specimen figured is stronger than that of most of the Venezuela specimens. The folding of the unfigured specimen is broader than that of its Grenada companion and of the Venezuela specimens as well, although similar to the latter in other respects. The shell is too dense for a successful count of the punctae.

Tichosina sp. 4

Plate 24, figure 1

Large, strongly suggesting Tertiary *Terebratula*; pentagonal in outline, widest near midvalve; sides well rounded; apical angle 85°. Valves unequal in depth, the dorsal valve being the shallower; beak large, foramen large and strongly labiate, permesothyridid. Anterior commissure gently uniplicate; lateral commissure straight. Color white. Both valves flattened anteriorly, the dorsal valve with a faint fold having a faint medial flattening or depression and the opposite valve with a faint folding opposite the depression in the dorsal fold. Interior damaged.

Measurements in mm	length	dorsal valve length	width	thickness	apical angle
550667	33.7	30.0	29.2	20.0	85°

Locality.—OREGON Station 4928.

Diagnosis.—Large, wide *Terebratula*-like *Tichosina* with large foramen.

Types.—Described specimen: USNM 550667.

Tichosina sp. 5

This is another large elongate oval specimen suggestive of *T.* sp. 4 but differing in having a less convex dorsal valve and less swollen dorsal umbonal region with anterior rounded and narrow. The specimen is 31 mm long, 25.4 mm wide and 17.3 mm thick.

Locality.—COMBAT Station 450.

Types.—Described specimen: USNM 550668.

Tichosina sp. 6

This is a species suggestive of *T.* sp. 1 but although of about the same size it is more triangular and has less inflated valves. Greatest width in the anterior third. The anterior commissure is broadly uniplicate. The dorsal valve is narrowly inflated posteriorly. The loop is entirely unlike that of *T.* sp. 1, as it is rather widely flaring anteriorly, which is unusual with the Caribbean Tichosinas. The loop of *T.* sp. 1 is narrow and has a narrow fold, whereas the loop of *T.* sp. 6 has only a broad flexure.

Measurements in mm	length	dorsal valve length	width	thickness	apical angle
550669	24.0	21.4	20.0	15.0?	76°

Locality.—OREGON Station 5015.

Diagnosis.—Subtriangular, inflated *Tichosina* with flaring loop.

Types.—Described specimen: USNM 550669.

Tichosina sp. 7.
Plate 3, figures 10-17

About medium size for the genus, elongate oval in outline with moderately rounded lateral margins and moderately convex anterior margin; posterior margins forming an angle of 65° to 80°. Lateral commissure anteriorly curved; anterior commissure broadly uniplicate, especially in the old adult; beak not strongly protruding; foramen fairly large, labiate and mesothyridid to submesothyridid. Color white and marked only by concentric lines; punctae numbering 140 per square millimeter.

Ventral valve moderately convex in lateral profile with maximum convexity at midvalve; anterior profile, a broad moderately convex dome with moderately steep slopes; umbonal region narrowly swollen, the swelling disappearing anteriorly where the valve flattens.

Dorsal valve evenly convex in lateral profile with the maximum convexity at about midvalve; anterior profile moderately and somewhat narrowly domed, more so than the ventral valve. Umbonal region swollen; flanks flattened, fairly long and steep.

Loop in length slightly less than one-third the valve length; outer hinge plates narrow but deep, margined by the crural bases which are flattened; crural process short and blunt, just posterior to the broad and broadly folded transverse ribbon.

Measurements in mm	length	dorsal valve length	width	thickness	apical angle
550611a	22.7	20.3	17.4	13.3	77°
550611b	21.6	18.8	16.9	13.3	77°
550611c	25.6	21.7	19.0	15.1	67°

Locality.—OREGON Station 2081.

Diagnosis.—Medium size, elongate oval *Tichosina.*

Types.—Figured specimens: USNM 550611a, b.

Comparison.—This species is about average size for the genus and is elongate oval in outline. It somewhat resembles *T. subtriangulata,* n. sp., but is more elongated, with a larger foramen, and is translucent white rather than yellowish. It is about the same size as *T. plicata* but it is not so strongly folded and has a broader more rounded anterior margin. *Tichosina* sp. 7 is smaller than *T. pillsburyae* and is not so strongly folded anteriorly as that species, but has a larger foramen. A larger series of specimens might establish this as a new species. It is of considerable interest because of its occurrence off the coast of Brazil, from which few brachiopods are known. South of Brazil in Argentine waters, brachiopods similar in appearance to *Tichosina* are referable to *Liothyrella* because they have a widely triangular loop.

Tichosina sp. 8

Plate 12, figures 5-7

Large for the genus, elongate oval in outline with the maximum width just anterior to midvalve; apical angle 70° to 80°. Anterior margin subtruncate. Ventral valve deeper than the dorsal valve. Anterior commissure gently uniplicate. Foramen large, strongly labiate, submesothyridid. Color pale, yellowish white, translucent. Punctae 76-79 per square millimeter.

Ventral valve strongly convex, narrowly swollen medially and with steep lateral slopes. Dorsal valve less convex than the ventral one and forming a broad dome in anterior profile. Loop narrow; transverse ribbon broad and only moderately folded medially; outer hinge plates long and deep. Cardinal process fairly large. Body wall strongly spiculated.

Measurements in mm.—(550546a); length 29.2; dorsal valve length 25.7; width 22.7, thickness 18.1; apical angle 80°.

Localities.—BLAKE Stations 155, 167.

Types.—Figured specimen: USNM 550546a.

Comparison.—The specimen figured here and another were taken with *Tichosina cubensis* (Pourtalès) according to museum records. Although approximately the same length as adult *T. cubensis,* it differs from that species in being much narrower, in having its anterior more truncated and in having a less sinuate lateral margin. Inside, the loop is wide and the transverse ribbon less angulated than that of *T. cubensis.* It seems unlikely that this is just a variant of *T. cubensis* because the characters are too well formed and different from that species. Larger collections of *T. cubensis* are needed in order to evaluate its variation. The punctae count is near that of the lower limit of *T. cubensis.*

Tichosina sp. 8 is also suggestive of *T. solida,* n. sp., but that species is more rounded laterally and anteriorly and its loop is narrower and the transverse ribbon more narrowly angulated.

Tichosina sp. 9

This is a thin-shelled species with small, strongly labiate foramen and strongly folded anterior commissure. Its loop is parallel-sided with a moderately broad transverse ribbon. The outer hinge plates are long and deep and the crura are short.

Locality.—G.704.

Types.—Described specimen: USNM 550623.

Discussion.—This species suggests *Erymnia angusta,* n. sp., in outline and small foramen but it does not have the loop welded to the floor of the valve. It occurs with *E. muralifera,* which is a much larger and rounder shell.

Erymnia Cooper, new genus

Externally like *Tichosina* with rounded, oval outline, unequally deep valves, the ventral valve deeper than the dorsal one and a uniplicate to rectimarginate anterior commissure. Color white to grayish white; surface marked by concentric lines and varices of growth and indistinct raised radial lines.

Ventral valve interior with a fairly long elevated pedicle collar. Muscle and pallial marks fairly strongly impressed. Dorsal valve interior with loop narrow but expanding slightly anteriorly, measuring about 1/4 to 1/5 the valve length. Cardinal process small, semicircular; socket ridges short, thin; outer hinge plates short, gently concave attached to the top (ventrad edge) of the crural bases which are broad like those of *Tichosina.* Attachment of outer hinge plates tapering to the crural process which is blunt and overhangs the transverse band; transverse ribbon longitudinally broad, medially folded and posteriorly fluted. Crural bases attached to valve floor by two nearly vertical plates which narrowly restrict the body wall.

Type-species.—*Erymnia muralifera* Cooper, n. sp.

Diagnosis.—Terebratulinae with loop attached to valve floor by partitions continuous with the crural base.

Derivation.—Greek *erymnos*=fence.

Comparison and discussion.—It was quite astonishing to discover in shells so similar in appearance to *Tichosina* ? *bartletti* (Dall) such a strong departure in the formation of the loop. The loop of *Erymnia* is similar to that of *Tichosina* in its construction, but the significant difference beside the attachment plates is the union of the inner edge of the outer hinge plates with the ventrad edge of the crural bases and the long tapering attachment which reaches almost to the crural process.

The attachment plates make for a very rigid loop but that structure in other species and genera seems fairly strong and solid. The posterior attachment plates restrict the body of the animal to a small chamber posterior to the posterior end of the transverse ribbon. In *Tichosina* the body wall expands anteriorly and laterally to cover much of the posterior third of the shell; the body cavity lateral and anterior to the loop is very shallow. In *Erymnia* the body wall is not spread out. The wall, as in most of these genera, is strongly spiculated.

This genus so far has only been found around the south side of Grand Bahama Island and the northeast side of the Straits of Florida. It is found in fairly deep water, usually in excess of 300 meters (1000 feet).

Erymnia angusta Cooper, new species

Plate 5, figures 16-24

Elongate oval or elliptical in outline, tapering at both ends and with the maximum width just anterior to the middle; sides somewhat bulging just anterior to midvalve; anterior narrowly rounded and posterolateral margins forming an angle of 87°. Beak small and narrow; foramen very small and strongly and narrowly labiate; permesothyridid; color opaque white; surface marked by concentric lines of growth. Anterior commissure narrowly uniplicate.

Ventral valve deeper than the dorsal valve, moderately convex in lateral profile but narrowly domed in anterior profile and with convex short sides. Median region and beak narrowly rounded and swollen, the swelling flattening anteriorly to form a steep anterior slope and short rounded tongue.

Dorsal valve flatly convex in lateral profile, with the greatest convexity in the umbonal region, anterior profile a broad, flattened dome; median region slightly inflated; fold originating in the anterior fifth, poorly defined and only slightly elevated.

Ventral valve interior with small teeth and short elevated pedicle collar; muscle and pallial marks well impressed.

Dorsal valve interior with loop equal in length to one-third the valve length; narrow and with a broad, sharply angular transverse ribbon. Outer hinge plates short and concave; crural bases forming a low elevation along the edge of the hinge plates. Crural bases with dorsad extensions which reach the valve floor at the posterior.

Measurements in mm	length	dorsal valve length	width	thickness	apical angle
550608a	29.9	23.7	20.1	15.8	87°

Locality.—SILVER BAY Station 3494.

Diagnosis.—Long slender, eliptical *Erymnia.*

Types.—Holotype: USNM 550608a; unfigured paratype: USNM 550608b.

Discussion.—This species differs strongly from *E. muralifera* in its external form, the latter being well rounded and subcircular in outline, the other elongate elliptical. The loop of *E. muralifera* is wider than that of *E. angusta* but it has subparallel sides and the sharply angulated broad transverse ribbon as in *E. angusta.* This species is like *Tichosina* sp. 9, but the loop of that species is not supported by struts.

Erymnia muralifera Cooper, new species
Plate 12, figures 3, 4; plate 13, figures 1-22; plate 14, figures 1-10.

Fairly large for a terebratulid, roundly oval to subcircular in outline; valve unequally convex, the ventral valve having the greater depth; maximum width at or anterior to midvalve; lateral commissure faintly S-shaped; anterior commissure broadly uniplicate. Apical angle near 90°. Beak low, narrow, truncated by a rather small foramen with strong, narrow lip; foramen permesothyridid. Color white to grayish white, surface lustrous. Ornament consisting of incremental growth lines only; punctae fine, crowded, numbering 75 per square millimeter.

Ventral valve fairly strongly convex in lateral profile with the maximum convexity at about midvalve; anterior profile strongly domed with long steep sides. Umbonal and median regions strongly swollen, but anterior slope flattened and extended into a short, broadly rounded tongue. Lateral slopes rounded and steep.

Dorsal valve less convex than the ventral valve and with the maximum convexity in the posterior half; anterior profile strongly domed but not so narrowly convex as the ventral valve in this view. Umbonal and median regions swollen, the swelling continued anteriorly to the front margin where the swelling produces a broad low fold. Flanks slightly depressed at the anterior but well rounded toward the posterior.

Ventral interior with small teeth, a long elevated pedicle collar and fairly strongly impressed muscle and pallial impressions. Dorsal valve interior with moderately strong and long socket ridges; outer hinge plates broad and only slightly concave, crural base not forming a partition along its inner, free edge, low and broad in profile and supported by a vertical plate along its dorsal edge which attached the loop to the valve floor. Crural processes blunt and rising to a crest just posterior to the transverse ribbon which is broad and fairly strongly arched medially.

Measurements in mm	length	dorsal valve length	width	thickness	apical angle
550521	29.0	26.2	25.4	19.3	89°
550520 (holotype)	29.1	26.3	24.5	16.5	88°
550523	26.5	24.0	23.8	15.1	96°

94

Locality.—G646, G691, G695, G704, G706; P991.

Diagnosis.—Roundly oval, rotund *Erymnia* with broadly uniplicate anterior commissure and small foramen.

Types.—Holotype: USNM 550520. Figured paratypes: USNM 550521a, 550522, 550523, 550578, 550624. Unfigured paratype: USNM 550521b.

Comparison and discussion.—Externally *E. muralifera* resembles *Tichosina cubensis* (Pourtalès) but is readily distinguished therefrom by its small foramen and the narrowly labiate beak. Its anterior commissure differs from that of *T. cubensis* in having a rather narrow dorsal curve whereas *T. cubensis* is usually rectimarginate but may be very broadly uniplicate. The interior will readily distinguish the two species. (See *E. angusta* for comparison with that species.)

Stenosarina Cooper, new genus

Elongate triangular in outline; valves convex but unequal in depth, the ventral valve having the greater depth; lateral commissure more or less sigmoidal in the adult; anterior commissure rectimarginate. Ventral valve narrowed at the posterior and with a strongly truncated beak; foramen usually small and variably labiate. Symphytium concave and usually only partially visible.

Pedicle valve with strong teeth buttressed by callus; muscle region long and narrow, often deeply impressed.

Dorsal valve with distinctive cardinalia, the socket ridges strong but the outer hinge plates fairly narrow, long and tapering; crural bases elevated along the inner margins of the outer hinge plates. Loop narrow, descending branches short and stout; crural processes broad and blunt, anterior of loop tapering and rounded; transverse ribbon very broad and strongly folded in the middle.

Type-species.—*Stenosarina angustata* Cooper, n. sp.

Diagnosis.—Narrowly elongate terebratulaceans having unequally convex valves, strongly truncated ventral beak, swollen dorsal umbo and a narrow, anteriorly tapering loop.

Comparison.—This genus is most like *Dallithyris* Muir-Wood (1959) in its external form. That genus has the ventral valve much deeper than the dorsal one and the same rounded triangular outline. The two differ in details of the loop. *Dallithyris* has a narrow loop with truncated anterolateral extremities but the crural bases do not form an elevated margin along the inside of the outer hinge plates. Furthermore, the crura are short and not flattened as in *Stenosarina.*

Tichosina cubensis (Pourtalès) is suggestive of *Dallithyris* and *Stenosarina* but it does not have the tapering loop of the latter, except as an aberrancy, and is provided with elevated crural bases along the edge of the outer hinge plates as in *Stenosarina.* Most species of *Tichosina* are quite unlike *Stenosarina* externally because their valves are usually subequal and they have a wider loop with angular anterolateral extremities. *Stenosarina* is externally somewhat similar to *Hispanirhynchia* Thomson and *Sphenarina* Cooper, but these have rhynchonellid cardinalia and no loop.

Species similar in external appearance to *Stenosarina* occur in the Cretaceous of Germany and the Eocene of Cuba. The loop of the former is not known; the Cuban species has a narrow loop but its distal extremity is not preserved. This is a rare genus in the Caribbean and Gulf of Mexico. Two species occur in both bodies of water.

Stenosarina angustata Cooper, new species

Plate 31, figures 26-33; plate 33, figures 7-11

Fairly large for the genus, thin and delicate shell, elongate subtriangular in outline; length greater than the width; valves unequally deep, the ventral valve having the greater depth. Sides gently rounded; anterolateral extremities narrowly rounded and anterior margin gently rounded. Posterolateral margins variable, forming an angle of 65° to 85°. Lateral commissure gently sigmoidal; anterior commissure rectimarginate. Beak very short and strongly truncated; foramen moderately large, labiate, permesothyridid; symphytium hidden. Color white to translucent; surface marked by concentric lines of growth; punctae dense, numbering 154 per square millimeter.

Ventral valve moderately convex in lateral profile and with the maximum curvature in the posterior half; anterior half flattened; anterior profile a fairly high, somewhat narrowed dome with long steep slopes. Umbonal region narrowly swollen the swelling gradually dying anteriorly in the flattening of the anterior half; flanks flattened and steep.

Dorsal valve strongly convex in lateral profile with the greatest convexity in the posterior half; anterior profile like that of the ventral valve but the convexity broader and the lateral slopes not so steep; umbonal region inflated, the swelling continued to beyond midvalve where the anterior slope is gently convex.

Ventral valve with small narrow teeth parallel to the side of the shell; pedicle collar short and anteriorly excavated. Muscle and pallial marks lightly impressed.

Dorsal interior with small transverse cardinal process; loop long and narrow with nearly parallel sides and equal in length to about 1/3 the valve length; socket ridges fairly stout; outer hinge plates long and deeply concave, margined by the elevated crural base; crura flat, crural processes bluntly pointed; transverse ribbon extending anterior to the crural processes and very broad and narrowly folded medianly. Lophophore with long lateral lobes.

Measurements in mm	length	dorsal valve length	maximum width	thickness	apical angle
550594 (holotype)	27.2	25.5	20.6	15.7	79°
550598	25.5	24.1	20.6	15.0	71°

Localities.—OREGON II Station 11133. SILVER BAY Station 1184.

Diagnosis.—Large, long, narrow *Stenosarina* with swollen dorsal umbo, strongly truncated ventral beak and narrow loop with broad transverse ribbon.

Types.—Holotype: USNM 550594. Unfigured paratype: USNM 550598. Figured specimen: USNM 550766.

Comparison and discussion.—At first glance this species suggests a narrow variant of *T. cubensis* (Pourtalès) because of its sigmoidal lateral commissure and the lidlike dorsal valve. It differs from *T. cubensis* in being much narrower, differently proportioned, having a smaller foramen and a long narrow loop. *Stenosarina angustata* is also externally very like *Dallithyris sphenoidea* (Philippi) from off Portugal, Morocco and the west coast of Africa. In fact, Davidson (1886: 12) included *T. cubensis* in that species. The West African specimens appear externally to be intermediate between *S. angustata* and *T. cubensis*. *Stenosarina angustata,* however, has a much shorter, more severely truncated anterior than that of *D. sphenoidea* and the American shells are more compressed laterally than those from Africa. The dissimilarity of the loop of *D. sphenoidea* to that of *T. cubensis* and *S. angustata* shows the external forms of each to be convergence toward an adaptive type.

Stenosarina nitens Cooper, new species

Plate 32, figures 16-21; plate 33, figures 1-6

About medium for the genus, rounded subtriangular to subpentagonal in outline; dorsal valve having less depth than the ventral valve; sides gently rounded; posterolateral margins forming an angle of 84-93°. Lateral commissure strongly sigmoidal; anterior commissure rectimarginate. Beak narrow, low, truncated and with a very small labiate foramen. Translucent white. Punctae numbering 132 per square millimeter.

Ventral valve with gently convex lateral profile but anterior profile somewhat narrowly convex. Umbonal region very narrowly swollen and having steep umbonal slopes; median region swollen but anterior flattened. Dorsal valve fairly strongly convex in lateral profile and broadly so in anterior view. Umbonal and median regions strongly swollen, so much so that the beak is almost hidden when the shell is viewed from the dorsal side.

Loop occupying about 1/4 the valve length, narrow, parallel-sided with blunt crural processes and strongly tapering distal end. Transverse ribbon broad and strongly folded medially.

Measurements in mm	length	dorsal valve length	maximum width	thickness	apical angle
550597	14.3	13.7	12.5	8.7	84°
550763 (holotype)	19.7	18.8	18.3	12.4	93°

Localities.—OREGON Station 5927; SILVER BAY Station 5181.

Diagnosis.—Medium-sized *Stenosarina* with low beak and small foramen.

97

Types.—Holotype: USNM 550763. Figured paratype: USNM 550597.

Comparison.—This species is most like *S. oregonate,* n. sp., in general form but is wider, rounder, more swollen at the dorsal umbo and with a smaller foramen. It also suggests the young of *Tichosina cubensis* (Pourtalès) but these of the same size as *S. nitens* have a broader, non-tapered loop quite unlike that of *S. nitens. Stenosarina angustata,* n. sp., is larger, longer and with a larger foramen than *S. nitens.*

Stenosarina oregonae Cooper, new species

Plate 31, figures 13-18

Of about medium size (25 mm), roundly oval to subpentagonal in outline; valves unequally deep, the ventral valve having the greater depth; lateral commissure gently sigmoidal; anterior commissure rectimarginate. Beak truncated, foramen large for the genus, moderately labiate. Surface smooth, color translucent white. Apical angle 73° to 80°. Punctae numbering 126 per square millimeter.

Ventral valve moderately convex in lateral profile but narrowly rounded in anterior profile. Beak region narrow and full but expanding anteriorly and laterally; anterior slope long; posterolateral margins very steep.

Dorsal valve unevenly convex with the umbonal region swollen; anterior profile broadly convex, more broadly so than that of the ventral valve; median region inflated with moderately steep slopes to the margins.

Ventral valve interior with narrow muscle region and with the muscle scars strongly impressed. Pallial trunks not strongly defined. Dorsal valve cardinalia as in the genus but the narrow loop broken.

Measurements in mm	length	dorsal valve length	maximum width	thickness	apical angle
550595a	24.4	23.3	20.8	15.0	80°
550595b	27.8	25.7	20.7	14.8	73°

Locality.—OREGON Station 4574.

Diagnosis.—Large *Stenosarina* with large foramen.

Types.—Holotype: USNM 550595a. Figured paratype: USNM 550595b.

Comparison.—This species is similar to *S. angustata* but differs in having a larger foramen, wider and larger beak, somewhat rounder anterior and deeper ventral valve. It is much larger than *S. parva,* n. sp. It is a larger, thicker shelled species than *S. nitens* and has a larger foramen and rounder anterior margin.

Stenosarina parva Cooper, new species

Plate 31, figures 19-25

Small for the genus, strongly and narrowly triangular in outline; posterolateral margins straight and forming an angle of 72°; anterior margin broadly rounded. Lateral commissure moderately sigmoidal; anterior commissure rectimarginate. Beak small and narrow, truncated; foramen small, narrowly labiate. Valves unequally convex, the ventral valve much deeper than the other. Yellowish white. Punctae numbering 94-99 per square millimeter.

Ventral valve very narrowly convex posteriorly; moderately convex in lateral profile but more highly convex in anterior profile. Umbonal region narrowly swollen, the swelling continuing to the anterior margin; lateral slopes very steep nearly to the anterior.

Dorsal valve strongly convex in lateral profile but broadly and only moderately convex in anterior profile; umbonal region swollen but lateral slopes short and moderately steep. Dorsal valve interior with very narrow loop measuring in length a third that of the valve. Transverse ribbon broad, strongly and narrowly folded.

Measurements in mm	length	dorsal valve length	maximum width	thickness	apical angle
550596	12.0	11.3	8.8	7.5	72°

Locality.—Johnson-Smithsonian Expedition Station 43.

Diagnosis.—Very small, narrow *Stenosarina.*

Types.—Holotype: USNM 550596.

Comparison.—This species is smaller and narrower than the others described herein. It is also more triangular, with straighter lateral margins and very small apical angle. Only one specimen is now known.

Genus *Dallithyris* Muir-Wood, 1959

Dallithyris murrayi Muir-Wood

Plate 33, figure 18

A figure of the loop of the type-species of *Dallithyris* is introduced for comparison with the loops of *Stenosarina* and *Tichosina.*

Types.—Figured hypotype: USNM 550332.

Superfamily CANCELLOTHYRIDACEA Cooper, 1973b

Family Cancellothyrididae Thomson, 1926

Subfamily Cancellothyridinae Thomson, 1926

Genus *Terebratulina* D'Orbigny, 1847

Terebratulina cailleti Crosse

Plate 25, figures 1-16; plate 28, figures 4-27

Terebratulina cailleti Crosse, 1865: 27, pl. 1, figs. 1-3.—Dall, 1871: 10.—Davidson, 1886: 26, pl. 5, figs. 41, 42.—Dall, 1920: 308.

This little species of *Terebratulina* is fairly common and widespread in the Caribbean and West Indian regions and the Gulf of Mexico (Cooper, 1954: 364). It is readily recognized by its narrowly elongate oval form and small size usually near or slightly in excess of 10 mm (ca. a half inch). Its color ranges from white in bleached specimens to pale yellowish. The shell is multicostellate with the primary, posterior costellae being stronger than the others which are intercalated anteriorly. Well preserved or young specimens have the costellae strongly beaded. Inside the dorsal valve is a short loop consisting of two stout crura bearing a ring consisting of the fused crural processes and a ventro-anteriorly projecting, flat transverse ribbon.

Localities.—G190, G304, G482, G692, G816; P584, P594, P600?, P658, P707, P708, P734, P737, P739, P838, P849, P861, P876, P903, P905, P929, P943, P954; OREGON 955, 1025, 4398, 4459, 4570, 5018; COMBAT 450. USGS Station D-19; west-southwest of Dry Tortugas at 183-457 meters (R. Cooper collection).

Types.—Figured hypotypes: USNM 550543, 550576, 550756a, b, d-h.

I have placed in this species two specimens from locality P600 which are larger than any specimens that hitherto have been referred to *T. cailleti*. One of these specimens is 16.2 mm long and 12.6 mm wide, and is somewhat more expanded anteriorly than the smaller *T. cailleti* specimens but this would be expected in a shell expanding by growth. A large specimen of *T. cailleti* in the University of Miami collection is 12.2 mm long and 8.7 mm wide. The width of a growth stage of the largest specimen in question taken at 12.2 mm anterior to the ventral beak is 8.6 mm which indicates an earlier stage of the large specimen to have been of about the same shape as the smaller one. Other similarities between these large specimens and the normal sized *T. cailleti* are the strong beading of the costellae, the median sulcus on the dorsal valve which creates a small dorsal fold of the anterior commissure, the narrow posterior swelling of the dorsal valve and the narrow and extremely stout loop. It is possible that the locality for these two large specimens was exceptionally favorable. The specimens are completely unlike the more northern *T. septentrionalis* (Couthouy). These specimens are also entirely unlike *T. latifrons* Dall, which attains nearly the same length as normal *T. cailleti* but is so much wider that additional growth to 16 mm would produce a still more strongly and widely triangular shell.

Terebratulina latifrons Dall
Plate 17, figures 1-13

Terebratulina cailleti var. *latifrons* Dall 1920: 309.

Small for the genus, subtriangular in outline with narrowly rounded sides and broadly rounded anterior margin; posterolateral margins nearly straight and forming an angle near 90°, slightly more or less. Valves of nearly equal depth, the dorsal valve slightly the deeper of the two. Anterior commissure broadly and angularly uniplicate. Beak short, narrow and with small but typically *Terebratulina* foramen; deltidial plates vestigial. Color white to yellowish. Surface somewhat fascicostellate, the primary costellae maintaining their identity from beak to anterior margin; secondary costellae appearing in three generations and tending to form bundles consisting of a primary and two or more subsidiary costellae. Costellae beaded.

Ventral valve gently convex in lateral profile, most convex in the posterior half; anterior profile very flatly convex and depressed medially by a narrow sulcus. Median region slightly inflated; sulcus slightly posterior of midvalve, deepening and widening anteriorly where it occupies about 2 mm of width. Flanks bounding sulcus gently convex; posterolateral slopes rounded and steep.

Dorsal valve evenly and gently convex in lateral view but forming a broad, subangular dome with narrowly, almost carinate middle and long flattened sloping sides. Umbonal region moderately swollen. Fold originating on the umbonal region, sharpening and widening anteriorly to the front margin where it is most pronounced.

Ventral valve interior with a thick and moderately long pedicle collar and thick elongated teeth. Muscles not strongly impressed. Dorsal valve interior with short, thin socket ridges protruding slightly posterior of the hinge; crura short and stout.

Measurements in mm	length	dorsal valve length	width	thickness	apical angle
314855a (Lectotype)	10.2	9.3	8.5	4.3	84°
314855b	10.6	9.1	9.3	5.2	82°
550517	10.1	9.1	9.9	4.3	89°
550518	11.5	10.0	11.4	6.0	91°

Localities.—Types: 35 fathoms, Barbados. G713, G947, G982, G983, G985, G986; P769, P1142; OREGON 2248, 5955.

Diagnosis.—Subtriangular *Terebratulina* with subcarinate fold and sulcus.

Types.—Lectotype: USNM 314855a. Figured hypotypes: USNM 550519, 550579a, b. Unfigured paratype: USNM 314855b.

Comparison and discussion.—*Terebratulina* with a fold and sulcus is a rarity in modern seas. *Terebratulina valdiviae* Blochmann from the Indian Ocean is such a species but it is so much larger and differently ornamented than the Caribbean species that no other comparison is necessary. *Terebratulina latifrons* is much wider and its fold and sulcus so pronounced that it is readily separable from *T. cailleti*, in which the fold and sulcus are indifferently developed. A species of *Terebratulina* from the Miocene of Cuba is very similar to *T. latifrons* but this has not yet been described.

The salmon tint to the inside of the valves of *T. latifrons*, noted by Dall in his one-line description, is not present in the Gerda and Pillsbury specimens and suggests a discoloration occurring after death.

Subfamily Eucalathinae Muir-Wood, 1965

Genus *Eucalathis* Fischer and Oehlert, 1890

Eucalathis cubensis Cooper, new species

Plate 16, figures 1-8

Eucalathis murrayi Dall (not Davidson), 1920: 323.

Shell small, widely triangular in outline, sides and anterior margin rounded; color white. Anterior commissure rectimarginate. Valves of unequal depth, the pedicle valve having the greater depth; maximum width anterior to midvalve; foramen large, deltidial plates absent. Surface costellate; costellae strongly and closely beaded, beading prominent and rounded; costellae about 30 in number, five of them primary, the others intercalated in two generations.

Pedicle valve gently convex in lateral profile and with the maximum convexity near midvalve; anterior profile moderately but broadly convex; lateral slopes gentle; median and umbonal regions swollen; anterior slope long and gentle. Median three costellae more prominent than others.

Brachial valve moderately convex in lateral profile and with the maximum convexity in the posterior half; anterior profile moderately convex but with long gentle lateral slopes; median region swollen; ears small. Median two costellae stronger than the others.

Brachial valve interior with stout and long crural plates; loop delicate and with a deep median angulation.

Measurements in mm	length	brachial valve length	width	thickness
Holotype	5.2	4.4	5.0	2.7

Types.—Holotype: USNM 110848.

Locality.—BLAKE Station 16

Discussion.—This species is characterized by the strongly beaded character of the costellae, length and width almost equal, and the long crural processes. Of described Recent species it may be compared to *E. trigona* (Jeffreys) but differs in its less attenuated form. It differs from *E. murrayi* (Davidson), to which it had been mistakenly referred, by its much more numerous costellae, the South Pacific form having only seventeen costellae. It is more like *E. floridensis,* n. sp., than either of the species mentioned but differs from that form in its rounded outline, more beaded costellae, different pattern of costellae and larger foramen.

102

Eucalathis floridensis Cooper, new species
Plate 16, figures 9-11

Small, triangular in outline, sides and anterior margins rounded; posterolateral margins straight. Color white, beak narrow, foramen large, deltidial plates vestigial. Anterior commissure rectimarginate; pedicle valve slightly deeper than the brachial valve; maximum width at about midvalve. Surface costellate, costellae strong, beaded in posterior half but nearly smooth anteriorly; costellae strong and narrowly rounded, about 28 in number; primary costellae about 10 in number and forming fascicles of two or three with the secondary costellae at the front margins.

Pedicle valve moderately convex in lateral profile and with the maximum convexity at midvalve; anterior profile broadly but gently convex; lateral slopes gentle; anterior slope long and gentle; posterior-median region moderately swollen. Median furrow deeper than others and forming a deep but narrow groove.

Brachial valve gently convex in lateral profile but with the maximum convexity in the posterior part; anterior profile broadly and gently convex. Ears small; median region swollen; anterior slope long and gentle. Midvalve marked by a costella more prominent than the others and opposite the deep groove of the pedicle valve.

Brachial valve with delicate loop and slender crura.

Measurements in mm	length	brachial valve length	width	thickness
Holotype	5.6	4.6	5.0	2.5

Horizon and locality.—Pourtalès Platform, Florida Keys at 366 meters.

Types.—Holotype USNM 107526.

Discussion.—The ornamentation and form of this shell are so distinctive that comparison to any known Recent species except *E. cubensis,* n. sp., is unnecessary. The outline and form are very close to those of *E. cubensis* as can be seen from a comparison of the measurements of the two. Other details however are so different as to make it impossible to place them in the same species. The beak of *E. floridensis* is more acute than that of *E. cubensis* but the striking differences are in the costellae. The ribbing of *E. floridensis* is much stronger than that of the Cuban shell and the beading is not so crowded and is confined to the posterior half, the front half being nearly smooth. Furthermore, the pattern formed by the costellae is quite different in the two species. The loops appear to have been different, that of *E. cubensis* being the stouter.

Family Chlidonophoridae Muir-Wood, 1959
Subfamily Chlidonophorinae Muir-Wood, 1959
Genus *Chlidonophora* Dall, 1903
Chlidonophora incerta (Davidson)
Plate 1, figure 1; plate 19, figures 8-22; plate 29, figures 1-13

Megerlia? *incerta* Davidson, 1878: 438; 1880: 49, pl. 11, figs. 17, 18.
Terebratulina? *incerta* (Davidson) Davidson, 1886: 38, pl. 6, figs. 23-25.
Chlidonophora incerta (Davidson) Dall, 1903: 1538; 1920: 322.—Cooper, 1954: 364.

This is a small and easily recognized genus because of its rounded form, multicostellate exterior and the unusual pedicle. The species resembles a round *Terebratulina* because the foramen is large and the deltidial plates are aborted or minute. The hinge is much wider than that of *Terebratulina* and a modest interarea is present. The cardinalia of *Chlidonophora* are almost exactly like those of *Terebratulina* except for the loop in which the crural processes do not join medially to form a ring. Instead the loop is like that of *Eucalathis* in which the transverse ribbon is extended anteroventrally to form a narrow angle.

Although these shell characters are quite distinctive, the pedicle is unique. In the large majority of specimens of *C. incerta,* the pedicle is widely frayed on its emergence from the foramen, and strongly resembles the byssus of some of the Bivalvia. The frayed ends attach to solid objects, but more often penetrate shells of *Globigerina* so common at the great depths at which this species occurs. More unusual is a long extension of the pedicle as a solid stem frayed at its distal extremity. Again the fibers of the pedicle penetrate foraminiferal shells. This development of the pedicle is rare in *Chlidonophora* from the Gulf of Mexico but it is usual in *Chlidonophora chuni* (Blochmann) from off the Maldives in the Indian Ocean.

Davidson (1880: 49) reports *Chlidonophora incerta* from the South Atlantic about 8.5° east of Fernando de Noronha, Brazil (1° 47′ N, 24° 26′ W). This is the southernmost occurrence so far recorded. It is found in deep water of the Caribbean Sea off the north coast of Panama, in the Colombian Basin south of Jamaica, off Havana, Cuba, west of Great Inagua, west of St. Vincent and off British Guiana. It is not known in the Atlantic north of the Caribbean, but occurs in the deeper waters of the Gulf of Mexico. The species ranges from 534 meters off Havana, Cuba (Dall, 1920: 322), the shallowest recorded occurrence, to 4798 meters off the northwest coast of Haiti. In the Gulf of Mexico it occurs down to 3843 meters.

Localities.—OREGON Station 2574; ALAMINOS Stations 4, 4c, 5, 12; ALBATROSS Station 2383; BLAKE Station 16; PILLSBURY Stations P325, P680, P871, P1138, P1266, P1397, P1401, P1402.

Types.—Figured hypotypes: USNM 550584a, 550586a, b, 550602a, b, 550663, 550757a-c.

104

Notozyga Cooper, new genus

Small, biconvex, valves nearly equal in depth; subquadrate to subpentagonal in outline. Lateral and anterior commissures straight, hinge wide and with fairly broad interareas on the ventral valve; delthyrium open. Deltidial plates small, narrow and marginal; apical callosity fairly long. Surface multicostellate; costellae in 2 generations, implanted and bifurcated; costellae beaded.

Ventral valve with large, transverse teeth; dental plates absent; other details not distinguishable.

Dorsal valve with deep sockets and high socket ridges extended beyond the posterior margin, the edges serving as a wide cardinal process; crural bases stout and wide passing into thick and angular crural processes. Transverse ribbon fairly broad, with a slight dorsad depression medially but the ribbon extending slightly posteriorly in uniting the two crura. Inner margins of both valves with recessed crenulations.

Type-species.—*Notozyga lowenstami* Cooper, new species. Named for its discoverer, Dr. Heinz A. Lowenstam.

Diagnosis.—Small Cancellothyrididae resembling *Eucalathis* but with transverse band of the loop not extending anterior to the crural processes.

Discussion and comparison.—At first glance this small shell suggests *Eucalathis* because of its pentagonal form and beaded costellae. It seems, however, to have a wider hinge and more prominent interareas on each side of the delthyrium. Its deltidial plates are differently developed than those of *Eucalathis*, which are short and triangular, whereas those of *Notozyga* are long and slender and margin the sides of the delthyrium. The chief difference between the two genera is in the loop. This structure in *Notozyga* is very stout and the transverse ribbon does not extend anteriorly to make a point as it does in all species of *Eucalathis*. *Notozyga* resembles *Gisilina*, *Rugia*, and *Meonia* all of Steinich (1963), but in each of these the transverse ribbon is pointed and the point is directed anteriorly.

Notozyga lowenstami Cooper, new species

Plate 16, figures 17-31

Small, subpentagonal to somewhat quadrate and with the beak forming an angle of about 90°. Strongly biconvex, sides nearly straight or sloping slightly posteriorly; anterior margin broadly rounded; cardinal extremities obtusely angular; commissures rectimarginate. Interarea broad, apsacline. Delthyrium wide and margined by narrow deltidial plates that extend nearly to the apex. Surface multicostellate, costellae appearing in two generations, the second about at midvalve and forming fascicles at the anterior; ventral valve marked medially by a narrow depression and dorsal valve with a corresponding elevated costella. Costellae coarsely beaded.

Interior as for the genus.

Measurements in mm	length	dorsal valve length	width	hinge width	thickness	apical angle
550591a (holotype)	2.8	2.1	2.6	1.7	1.6	92°
550591b	2.8	2.0	2.2	1.5	1.5	88°
550591c	2.6	2.1	2.1	1.8	1.5	99°
550591e	1.7	1.3	1.6	1.0	0.8	94°

Locality.—California Institute of Technology station 1484.

Diagnosis.—Strongly biconvex, fascicostellate *Notozyga.*

Types.—Holotype USNM 550591a. Figured and measured paratypes: USNM 550591b, c, e; unfigured paratype: USNM 550591d.

Discussion.—The fasciculate ornament and wide hinge distinguish this species externally from species of *Eucalathis.* The loop is completely unlike that of *Eucalathis, Chlidonophora* and the Cancellothyrididae. No other species is known.

Suborder TEREBRATELLIDINA Muir-Wood, 1955

Superfamily TEREBRATELLACEA King 1850

Family Megathyrididae Dall, 1870

Genus *Argyrotheca* Dall, 1900

This genus is very prolific in species, not only in the Recent fauna but also in the Tertiary and back into the Cretaceous. Today it is known in the Mediterranean and adjacent Atlantic, in the Indian Ocean and Red Sea, and in the Pacific where it is usually smaller than normal. Although fairly common in Mediterranean waters, it is generally rare elsewhere except in the Caribbean region, where it is common and is represented by several species already named and very likely many more yet to be described.

Argyrotheca has a long history in the Caribbean and West Indian region. It occurs in the Eocene of Cuba and the Gulf Coast of the United States. It is common in the Miocene of Cuba, less common in the Miocene of the United States. At present in the Caribbean it is represented by six described species to which two are added here. The collections made by the PILLSBURY contain a number of specimens not referable to the six already described species. Furthermore, the national collection contains many specimens from the Tertiary of the United States and Caribbean region not referable to any described species. The genus is evidently vigorous, adapting readily to new conditions especially in the area under discussion. In the Pacific, the majority of specimens so far found are minute but probably adult. One large species, *A. lowei* Hertlein and Grant, occurs in the Gulf of California.

Most members of the genus seem to prefer fairly shallow water and this is true of the known species from the Caribbean. The deepest record is of *A. barrettiana* (Davidson) from 805 fathoms (= 1473 meters) recorded by Dall (1920: 329).

In addition to preferring shallow water, many species of *Argyrotheca* occur attached to the underside of coral or coralline structures. They also occupy grottos and dark secluded, quiet places among the corralines (Jackson, Goreau and Hartman, 1971). Specimens are commonly so closely attached to their substrate that the ventral beak and often part of the dorsal valve are misshapen and corroded. The pedicle has "rootlets" that actually "eat" their way into the limy substrate, apparently by solution. Often in breaking a specimen loose from the substrate, part of the coralline surface comes off with the brachiopod.

Argyrotheca barrettiana (Davidson)

Plate 22, figures 9-21; plate 23, figures 6, 7; plate 32, figures 22-32

Argiope barrettiana Davidson, 1866 (February): 103, pl. 12, fig. 3.
A. antillarum Crosse and Fischer, 1866 (July): 270, pl. 8, fig. 7.
Cistella Barrettiana Davidson, 1887: 145, pl. 22, figs. 1, 2.
Argyrotheca barrettiana (Davidson) Dall, 1920: 329.

This is the most colorful brachiopod found in the Caribbean region, with its straw yellow ribs and scarlet interspaces. It is widely distributed in this region and has been taken from the Gulf of Mexico and as far south as Rio de Janeiro, Brazil (Davidson, 1887: 146). Unfortunately for ease of identification, this is not the only *Argyrotheca* with red or scarlet bands in the interspaces. The others are usually smaller or have salmon colored costae. Nevertheless, the distinction between red-banded *A. rubrotincta* Dall and the young of *A. barrettiana* is difficult.

Argyrotheca barrettiana is large for the genus, multicostate, the costae numbering up to 20 or slightly more, and with rectangular or obtuse cardinal extremities in the adult. In the young, the hinge may be wider than midwidth and the cardinal extremities acute or strongly pointed. The outline of the young is often transversely rectangular rather than nearly square or longitudinally rectangular. The costae of *A. barrettiana* are increased anteriorly by intercalation, the first intercalated series appearing near the beak and then in two generations anterior to that, the last generation being short and marginal. As usual in the genus, the folding is opposite so that the sulcus on each valve produces a median indentation.

Argyrotheca barrettiana generally favors waters shallower than 200 meters but has been taken from 1473 meters (= 805 fathoms).

Localities.—G533, G636, G983, G984, G985, G986; P330, P389, P409, P415?, P421, P422, P630, P924, P991, P1157, P1386, P1434; BLAKE 45.

Types.—Figured hypotypes: USNM 64229a-e, 550604a, b, 550605, 550747a.

Argyrotheca bermudana Dall

Plate 3, figures 1-5

Argyrotheca bermudana Dall, 1911: 86; 1920: 327.
Cistella cistellula Verrill (not Searles Wood) 1900: 592, pl. 70, fig. 7.

This is a small and distinctive species, thin-shelled, nearly smooth, with wide foramen and apsacline interarea. Its cardinal extremities are obtuse and the anterior margin is deeply indented medially by a sulcus on each valve to produce a strong bilobation. For immediate and easy recognition its color pattern is unlike that of any other Caribbean *Argyrotheca*. Like *A. barrettiana* it has scarlet stripes but these are roughly at right angles to the margins, thus the bands on the posterior half are nearly parallel to the hinge and those anteriorly are radial. The shell between the bands is white, or in liquid, nearly colorless.

Localities.—P855, P856.

Types.—Figured hypotype: 550531.

Discussion.—Logan (1974) describes the habitat and ecology of this species and notes that it is cryptic in habit and occurs on the undersides of foliaceous hermotypic corals. Distribution is controlled by the abundance of corals and their growth form. He notes differences of occurrence between Bermudian and Caribbean forms (Jackson *et al.*, 1971), the former less common, absent from internal reef cavities and not associated with sclerosponges.

Argyrotheca crassa Cooper, new species

Plate 25, figures 17-22

Argyrotheca lutea Dall (part), 1920: 320 (all Barbados citations).

Fairly large for the genus, thin-shelled, length and width nearly equal, sub-pentagonal in outline; sides slightly oblique and slightly rounded; anterior margin narrowly emarginate and serrate. Apical angle approximately 112°. Hinge narrower than the midwidth. Valves unequally convex, the ventral valve the deeper and more convex; surface costate, costae strong, broad and rounded, expanding anteriorly and numbering four or five on a side. Spaces between costae about equal to the width of a costa; additional costae intercalated anteriorly, narrow and often indistinct. Both valves with a prominent median sulcus. Color pale yellowish white.

Ventral valve moderately and evenly convex in lateral profile, broadly convex in anterior profile, the median crest indented by the median sulcus. Interarea fairly well developed, apsacline. Sulcus originating at the beak, rapidly expanding to the margin where it occupies a little less than half the shell width. Sulcus usually with two intercalated costae that extend from the anterior margin for variable distances usually not reaching the beak but occasionally attaining that point.

Dorsal valve gently to moderately convex in lateral profile; anterior profile varying from nearly concave to moderately convex; lateral slopes in anterior profile short and steep. Sulcus originating at the beak, broadening rapidly anteriorly, somewhat angular and deep, indenting the anterior margin where it meets the ventral sulcus. Intercalated costae introduced as in the ventral sulcus.

Ventral valve interior having narrow but thin and rather delicate teeth; apical plate or pedicle collar moderately long, anteriorly excavated and supported by a thin median septum that rises to a rounded crest posterior to midvalve; anterior slope of septum gradual to about midvalve where it joins the ridge produced by the deep median sulcus. No thickening of the muscle region.

Dorsal valve with short and narrow socket ridges; crural processes moderately long, bluntly pointed. Loop extending to midvalve where it joins the median septum. Area surrounded by loop moderately thickened for attachment of the adductor muscles. Median septum fan-shaped in lateral view, strongly crested at or slightly posterior to midvalve. Posterior slope of septum steep, the extension to the posterior part low and inconspicuous. Anterior slope of median septum variable, usually strongly serrated and provided with four or five narrowly rounded projections.

Measurements in mm	length	dorsal valve length	hinge width	mid-width	thickness	apical angle
550581 (holotype)	6.4	5.4	6.0	7.1	3.3	112°
314880a	7.6	6.6	5.0	8.2	4.1	113°
314878a	7.0	6.0	4.7	7.1	4.2	108°
314884a	6.9	5.6	5.1	7.7	3.8	109°

Localities.—P857. Several stations off Barbados.

Diagnosis.—Pale yellow, strongly costate *Argyrotheca* having a deep sulcus in each valve, and a strongly serrate and emarginate anterior margin.

Types.—Holotype USNM 550581. Unfigured measured paratypes: USNM 314878a; 314880a; 314884a; unfigured paratypes: USNM 314878b, c, 314880b, c, 314884b-h.

Comparison and discussion.—The pale yellow color of the entire shell will distinguish this species from *A. barrettiana* (Davidson), *A. johnsoni* Cooper and *A. rubrocostata,* n. sp., all of which have scarlet or red in some part of their shell. It is much larger and differently shaped than *A. schrammi* (Crosse and Fisher). The species nearest to *A. crassa,* with which it has been confused, is *A. lutea* (Dall). It is approximately the same size or a little larger than the known specimens of Dall's species and is much more strongly costate, has more prominent median sulci and is strongly serrate marginally except in very old specimens.

Typical *Argyrotheca lutea* (Dall) seems to be confined to the Gulf of Mexico and Western Caribbean as shown by specimens in the national collection, while *A. crassa* is developed in the eastern part of the Caribbean region. Fifteen lots from around Barbados are all referable to *A. crassa.* A single small specimen (USNM 62342) referred to *A. lutea* by Dall 1920, p. 320) off Rio de Janeiro is strongly costate and has the characters of *A. crassa.*

Argyrotheca lutea is smaller, differently shaped and with different costae when compared to *A.* sp. 1 and *A.* sp. 2, although all three are similarly colored.

109

Argyrotheca hewatti Cooper, new species

Plate 32, figures 1-15

Medium size for the genus, ventral valve hemipyramidal and dorsal valve flat to slightly convex; length and width nearly equal to slightly wider than long, depending on length of interarea and beak. Interarea strongly deformed, usually strongly apsacline but shape and inclination dependent on nature of substrate. Sides nearly straight; anterior margin strongly rounded. Anterior commissure with a faint ventrad wave. Hinge straight, generally forming the widest part and extended into small ears. Deltidial plates small, variable, elevated obliquely to interarea. Foramen large, occupying a half to 3/5 the hinge line. Costate, costae varying from 9 to 10 primary ribs with a generation bifurcated or intercalated at the anterior. Color of dried specimens pinkish overall but interspaces reddish and costae grayish white to grayish yellow.

Ventral valve convex in lateral profile but strongly domed in anterior profile; median region swollen. Interior with a slender median septum extending to about midvalve and supporting a short apical plate; median septum at junction with valve floor marked by a small pit. Teeth large; muscle region strongly thickened.

Dorsal valve flat to slightly convex in lateral profile and the same in anterior view. Sulcus narrow and indistinct, usually occupied by a costa. Ears slightly depressed. Interarea very short.

Dorsal valve interior with short stout loop; descending branches narrowly curved and meeting valve floor at about midvalve; anterior or ascending processes joining the median septum about 2/3 length of valve from beak: median septum stout and elevated at the anterior near valve edge; anterior slope of septum not serrated; proximal end of septum thickened. Region about the septum and the adductor field enormously thickened, the anterior thickening forming a V-shaped platform extending from the end of the septum to the valve floor and burying the median part of the septum.

Measurements in mm	length	dorsal valve length	midwidth	hinge width	thickness
550739a	4.9	3.2	5.1	5.3	2.9
550739b	5.1	3.8	5.3	5.5	3.0
550739c	5.1	4.2	6.0	6.2	3.7
550739d (holotype)	4.2	3.5	5.3	5.4	2.7

Locality.—Gulf of Mexico, 150 miles southwest of Sabine Pass, Texas.

Diagnosis.—Medium sized, reddish *Argyrotheca* with length and width nearly equal and with enormously thickened interior structures in the dorsal valve.

Types.—Holotype: USNM 550739d. Figured paratypes: USNM 550739a-c, e, f.

110

Comparison and discussion.—This species at once suggests the reddish species: *A. barrettiana* (Davidson), *A. johnsoni* Cooper and *A. rubrocostata* Cooper. *Argyrotheca barrettiana* is a larger and more finely costate species as well as being conspicuously marked by scarlet bands in the spaces between the yellowish costae. *Argyrotheca johnsoni* and *A. rubrocostata* have red costae and salmon interspaces and both are much larger and more transverse than *A. hewatti*. They are also more strongly costate than *A. hewatti*. None of the three species mentioned has the extravagant internal thickenings that characterize *A. hewatti*.

The species is named after Dr. Willis G. Hewatt, who presented the specimens to the Smithsonian Institution.

Argyrotheca johnsoni Cooper

Plate 1, figures 2-7

Argyrotheca johnsoni Cooper, 1934: 2, pl. 2, figs. 1-12.

This species is very similar to *A. rubrocostata* but is much more strongly ribbed, although its color is like that of the PILLSBURY specimens. *Argyrotheca johnsoni* was first found off the Dominican Republic, but since then Jeremy Jackson (1971) has collected numerous specimens on the north side of Jamaica in shallow water. See *A. rubrocostata* for comparison.

Localities.—P1153, P1421. Johnson-Smithsonian Station 52. Discovery Bay (J. Jackson). West Bull, Jamaica (H. Lowenstam).

Types.—Holotype: USNM 431003. Paratypes: USNM 431003a-c.

Argyrotheca lutea (Dall)

Plate 24, figures 12-28

Cistella lutea Dall, 1871: 20, pl. 1, figs. 5, 5a, pl. 2, figs. 4-8.—Davidson (part), 1887: 142, pl. 23, fig. ?5, 6.
Argyrotheca lutea (Dall), 1920 (part): 329.

About medium size for the genus, a uniform pale yellowish white; rectangular in outline; width slightly greater than the length; hinge narrower than the maximum width at about midvalve; anterior margin broadly rounded but only slightly indented medially. Anterior commissure deflected slightly in a ventral direction. Interarea well developed, apsacline. Surface costate, costae strong and rounded, separated by spaces about equal to the width of the costae; intercalations frequent, occurring from midvalve to anterior margin. Anterior margin faintly serrate or nearly smooth.

Ventral valve moderately convex in lateral profile, but somewhat narrowly rounded medially when viewed from the anterior. Sulcus originating at the beak, widening anteriorly, moderately deep at the front margin but indenting the margin only slightly; sulcus occupied by two costae, the length of which is variable.

Dorsal valve gently convex in lateral profile and broadly and gently convex in anterior view. Sulcus originating at the beak, widening rapidly anteriorly and broad and shallow at the front; sulcus occupied by two to four intercalated costae.

Ventral valve interior with elongated and flattened teeth; apical plate short and anteriorly excavated, supported by a short slender septum which thickens anteriorly but is not strongly elevated; septum indented by three pits between midvalve and anterior.

Dorsal valve with strongly elevated septum anteriorly, the crest at about midvalve becoming low and inconspicuous posteriorly between midvalve and the apex. Anterior of septum marked by three strong projections or serrations. Loop attaching to the median septum at midvalve; interior of valve not greatly thickened.

Measurements in mm	length	dorsal valve length	hinge width	mid-width	thickness	apical angle
110963a (lectotype)	?	5.1	4.7	6.6	?	?
110963b	5.5	?	5.2	6.8	?	109°
110963c	4.8	4.0	5.1	5.8	3.9	109°
110963d	5.6	?	6.2	6.8	?	112°
32924a	4.6	4.1	5.2	6.0	3.0	115°
32924b	5.3	4.8	5.8	6.6	3.8	109°

Localities.—Tortugas Pass at 30-43 fathoms (= 55-79 meters); off Havana, Cuba at 80-127 fathoms (= 146-234 meters); Ensenada de Cochinos, Cuba at 100-150 fathoms (= 183-275 meters).

Diagnosis.—Medium sized, uniformly yellowish *Argyrotheca* having length about 0.8 the width, strong costae and wide but shallow sulci not strongly indenting the margin or producing strong serration.

Types.—Lectotype USNM 110963a. Paratypes USNM 110963b, c. Figured hypotypes: USNM 32924a, b.

Comparison and discussion.—This species is smaller and less strongly costate than *Argyrotheca crassa,* n. sp., nor does it have the very deep sulci, strong anterior emargination and strong serration of *A. crassa,* which is best developed in the Antillean region. *Argyrotheca lutea* in color resembles *A.* sp. 1 and sp. 2, but there the resemblance ends. It is smaller, narrower, without serrate margins and with less prominent median sulci.

In designating a type locality for *A. lutea* Dall included specimens from 30 and 43 fathoms but in the collection only the specimens from 43 fathoms are labelled "Fig'd types." It is from the latter lot that Dall figured the one dorsal valve that constitutes his only figure of the shell. Other figures are of fleshy parts. In the "type" lot (USNM 110963) are a complete specimen and 4 separate valves, two ventral and two dorsal valves. Two are pairs (USNM 110963b) but the other two do not belong together. The dorsal valve of this odd pair is undoubtedly the specimen figured by Dall (1871, pl. 1, figs. 5, 5a) and must be chosen as lectotype.

This species is illustrated by Davidson (1887, pl. 23, figs. 5, 5a, and 6), but his figures 5 and 5a of the exterior are so highly exaggerated that they do not properly depict the species.

This species is rare and has been widely misidentified. Its report from the Barbados is incorrect, those specimens now being referred to *A. crassa*, n. sp. The occurrence off Rio de Janeiro appears to be a misidentification. Outside of the Gulf of Mexico, it is correctly identified from Ensenada de Cochinos on the south side of Cuba and from off Havana on the north side of Cuba.

Argyrotheca rubrocostata Cooper, new species

Plate 21, figures 12-27; text figure 8B

Fairly large for the genus, narrowly, transversely rectangular in outline, with the hinge equal to or wider than the midwidth. Cardinal extremities acute to obtuse; overall color reddish with the costae red and the interspaces salmon colored. Surface costate, costae opposite and forming a serrate margin; costae numbering six or seven on a side but the costae splitting near the margin. Interarea broad, curved, strongly apsacline to procline; foramen only moderately large; anterior commissure broadly and gently sulcate; profile nearly plano-convex with the ventral valve having the greater depth.

Ventral valve very gently convex in lateral profile and broadly domed in anterior profile; beak usually corroded; median region swollen, the swelling extending laterally to the flanks. Sulcus indistinct, widening anteriorly but narrow throughout its extent. Flanks sloping strongly to the sides.

Dorsal valve flat to faintly convex in lateral profile; anterior profile nearly flat to broadly concave; sulcus variable from narrow to fairly broad and with an intercalated costa anteriorly. Flanks flat or gently concave.

Ventral interior with small deltidial plates, short pedicle collar, thin median septum highest in the posterior about 2 mm anterior to the beak, then with a steep slope to a low ridge extending on the valve floor to within 2 mm of the anterior margin. Dorsal valve with broad crural processes and lateral lamellae attaching to valve floor near the margin at midvalve. Septum thick, strongly elevated anteriorly and with anterior edge marked by three serrations.

Muscles. — Musculature like that of *Megerlia* with large and flabellate ventral adjustor muscles and small, feeble diductors between them. Dorsal adjustors large, attached to pedicle in ventral valve but just anterior to notothyrial platform and each side of the median septum. Adductors as usual in articulate brachiopods.

Pedicle.—Short, fleshy, with dark brown, rootlike holdfasts that are capable of dissolving their way into limy substrate. Pedicle not attached to shell but forming the median part of the integument covering the delthyrium and overlying the place of attachment of the ventral and dorsal adjustors.

113

Measurements in mm	length	dorsal valve length	mid-width	width	hinge thickness
550529a	6.4	5.5	8.6	8.3	3.8
550530a	7.6	6.0	9.4	9.4	3.6
550530b	4.7	4.1	8.3	8.7	2.8

Localities.—P331, P419, P420, P629, P630. Carrie-Bow Cay, British Hondruas, latitude 16° 48′ N, longtiude 88° 05′ W, at 90 feet.

Diagnosis.—Salmon red *Argyrotheca* with the costae salmon red to red and the interspaces pale salmon color.

Types.—Holotype: USNM 550529a. Figured paratypes: USNM 550529b, 550530a, b.

Comparison and discussion.—This species is like *A. barrettiana* in having a bicolored shell but the colors are reversed, the scarlet of *A. barrettiana* occupying the troughs between the costae and the costae having a yellow color. The red of *A. rubrocostata* colors the costae and the troughs are a much lighter salmon color. Otherwise the two species are similar. Both are very variable in all their features, but *A. barrettiana* tends to have more numerous and finer costae. *Argyrotheca rubrocostata* tends to be wider in most of the specimens than *A. barrettiana* and more strongly costate but the young of both species are wider than long. In these respects *A. rubrocostata* is closer to *A. johnsoni* Cooper.

Argyrotheca johnsoni is colored like *A. rubrocostata* and the ribbing of the two is similar. That of *A. johnsoni,* however, is much stronger than that of the PILLSBURY specimens and this is an important difference (compare plates 1 and 21). The anterior scalloping of *A. johnsoni* is stronger than that of *A. rubrocostata* because of the coarser ribbing. The sulcus of the ventral valve of *A. rubrocostata* is usually narrow but is variable; that of *A. johnsoni* is also variable but is usually wide and commonly occupied by a median costa. The latter feature is uncommon in the PILLSBURY material. Bigger collections of both species are needed to fix these variations.

Muscles.—One exceptionally well preserved specimen (USNM 550530b) shows the muscles in fine detail. Their arrangement is not like that of most of the standardized terebratulids but is exactly like that of *Megerlia* (see text figure 8). The musculature of *Argyrotheca* and *Megerlia* is notable for the great development of the pedicle or adjustor muscles, which are far larger and more powerful than the diductors or adductors. The diductor muscles are intimately connected with the integument bearing the pedicle.

When viewing the interior of the ventral valve the large muscles so suggestive of diductors in the more usual terebratulid arrangement are seen to be the ventral pedicle or adjustor muscles. They form a large triangular patch on each side of the median line and are attached to the pedicle or the skin just beneath the pedicle and also to the unusually large dorsal adjustor muscles. In the dorsal valve these latter muscles are attached on each side of the median septum and on the sides of the septum just anterior to the notothyrial platform. In the ventral valve the dorsal adjustors attach to the skin under the pedicle and to the flattened ends of the ventral adjustors. The diductor muscles are small and are located between and slightly behind the ventral adjustors. They are conical at their base but thin to flat ribbons that extend posterodorsally to attach to the inner or anteroventral side of the integument that bears the pedicle. These diductor ribbons broaden along the inside of the integument and are attached to the posterior median edge of the dorsal valve with the integument. In *Megerlia* the diductors are also attached to the integument but do not extend to the edge of the dorsal valve (see figure 8). Thus the integument in these two genera must share in the process of opening the valves. The adductor muscles are attached near midvalve in both shells, two muscles in the ventral valve, each dividing to produce four in the dorsal valve, as usual in articulate brachiopods.

Pedicle.—The pedicle of this species of *Argyrotheca* is not like that of the usual terebratulid. It is soft, fleshy and round, not elongated but flattened and dislike, located in the middle of the integument covering the delthyrium. It is nowhere attached to the shell but seems an integral part of the integument. Within the soft fleshy mass of the pedicle are scattered dark brown, rootlike fibers whose distal ends have the ability to dissolve calcium carbonate and thus root the *Argyrotheca* to the coral, alga or other calcareous substrate (see plate 23, figures 6, 7). This is the reason why so many dried specimens from the Caribbean have part of the substrate clinging to the apical region, like lumps of soil caught by uprooted plants.

Action of the muscles.—*Argyrotheca* lives tightly attached to the substrate, so tightly that the posterior of both valves may be distorted. The shells are so strongly attached that they cannot turn on the pedicle nor can the dorsal valve be opened unless the animal is lifted above the substrate. It is here postulated that the great development of the adjustor or pedicle muscles and the degeneration of the diductors is to lift the shell above the substrate so that its shell can gape (Shipley, 1883: 504). Concerted action of all the adjustors and diductors will pull the integument into the valve, thereby stretching the pedicle. This movement will cause the valves to elevate sufficiently to permit them to gape. Relaxation of these muscles will allow the shell to settle to its previous position and action of the adductors will close the valves (Cooper 1973A: 12, 13).

Argyrotheca rubrotincta (Dall)

Plate 24, figures 2-11

Cistella (?*schrammi* var.) *rubrotincta* Dall, 1871: 19, pl. 1, figs. 6, 6a.
C. barrettiana var. *rubrotincta* Dall, 1886: 203.
C. barrettiana Davidson (part), 1887: 145, pl. 23, figs. 1, 2 only.
Argyrotheca schrammi (Crosse and Fischer) part, Dall, 1920: 330.

This is a small, transverse species usually having a strongly scalloped anterior margin in young and medium-sized specimens but with the margin tending to be evened in old specimens. The valves are strongly un-equal, the ventral valve being swollen and the dorsal valve nearly flat to gently concave. The coloring is distinctive, the ribs being pale yellow and the interspaces scarlet as in *Argyrotheca barrettiana* (Davidson). Young specimens have four strong, rounded costae on a side, and while older specimens have anterior intercalations, these are quite irregular.

Inside the ventral valve the septum is thin and delicate posteriorly but becomes thicker anteriorly. The anterior portion is marked by a pit near midvalve which is an accommodation to the sharp end of the dorsal valve

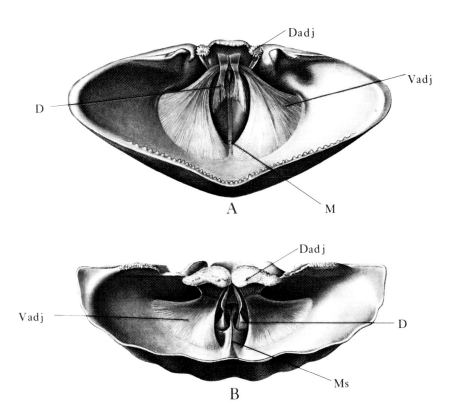

FIGURE 8. Interior of the ventral valve of: A: *Megerlia truncata* (Linnaeus) and B. *Argyrotheca rubrocostata,* n. sp. showing the similarity of the muscle arrangement in the two genera. Note particularly the great development of the pedicle adjustor muscles, the strong dorsal adjustors and the feeble diductor muscles that are attached to the integument surrounding the pedicle.
D: Diductors; Dadj: dorsal adjustors; M: myophragm; Ms: median septum; Vadj: ventral adjustors. A: ca x 10; B: ca x 17.

Drawings by Lawrence B. Isham, visual information specialist.

median septum. The median septum of the dorsal valve is thin and very high, rising to a crest and sharp point, colored red, near midvalve. The anterior slope of the septum is smooth and has a shallow depression just anterior to the crest. The posterior slope of the septum is steep. The muscle field is fairly strongly thickened.

Measurements in mm	length	dorsal valve length	hinge width	mid-width	thickness
110962a (lectotype)	3.7	3.1	4.5	4.3	2.7
110962b	3.1	2.2	4.7	4.6	2.0

Locality.—West of [Dry] Tortugas, 30-43 fathoms (= 55-79 meters). Florida Keys. P854?.

Diagnosis.—Small, transverse, serrate *Argyrotheca* with yellow costae and scarlet interspaces.

Types.—Lectotype: USNM 110962a. Paratypes: USNM 110962b-d. Figured hypotype: USNM 82923a.

Comparison and discussion.—Small size and the distinctive red and yellow color distinguish this species from described forms except *A. barrettiana* (Davidson) that is similarly colored. The latter is much larger, more numerously costated and its young entirely different from adult *A. rubrotincta*. Only a few specimens are known from the Gulf of Mexico.

Many specimens from Barbados in the Henderson collection in the National Museum of Natural History were identified as *A. rubrotincta* by Dall. These are very uniform with strongly serrated anterior margins. These specimens are consistent in their characters and conform fairly well to the young *A. rubrotincta* but not to the adult. Their relationship to type *A. rubrotincta* is uncertain.

Davidson's figures (1887, pl. 23, figs. 1, 1a) of Dall's type specimen are very inaccurate and give no true conception of the species. The profile of the median septum of the dorsal valve is likewise inaccurate. The large majority of specimens of this species have not attained the size of the lectotype and are usually strongly serrate as noted above.

Dall failed to designate a type specimen for his "variety." I therefore select specimen USNM 110962a (Dall, 1871, pl. 1, figs. 6, 6a). The two valves of this specimen have been separated to reveal the interior.

Argyrotheca schrammi (Crosse and Fischer)

Plate 16, figures 12-16

Argiope schrammi Crosse and Fischer, 1866: 269, pl. viii, fig. 6.
Cistella schrammi (Crosse and Fischer) Davidson, 1887: 148, pl. 22, figs. 3, 4.
Argyrotheca schrammi (Crosse and Fischer) Dall, 1920: 329.

This is a small species tinted reddish or salmon with the color evenly suffused over the whole shell. It is wider than long, with acute lateral extremities, strongly bilobate and serrated anterior margin. The sides are occupied by three costae with a trace of a costa in the median sulci. The species is very poorly known.

Locality.—NGS-MRF-33.

Types.—Holotype: USNM 550601.

Discussion.—This species is similar in size and shape to *A. rubrotincta* (Dall) but the coloration is different, the costation stronger and less numerous and the median sulcus much wider. This is not a well known species and a good series is needed to make it understood. The illustrated specimen was taken between 14 and 15 meters. Crosse and Fischer (1866) report that the type specimen was dredged between 200 and 250 meters.

Argyrotheca woodwardiana (Davidson)

Argiope Woodwardiana Davidson, 1866: 103, pl. 12, fig. 4.
Cistella Woodwardiana Davidson, 1886: 140, pl. 22, figs. 7-7c.

This is a very small species, pentagonal in outline with a long beak having a whitish yellow shell with a few red patches arranged in interrupted lines radiating from the beak. The shell is also marked with numerous concentric lines of growth. It was taken from 60 fathoms (110 meters). It has not been identified since its discovery off the northeast coast of Jamaica.

A number of small, smooth and red-splotched Argyrothecas were taken by the research vessels of the University of Miami but none of them

Argyrotheca sp. 1

Plate 23, figures 25-29

Large for the genus, pale yellow; transversely but broadly elliptical in outline; sides and anterior rounded; hinge narrower than the widest part which is at midwidth. Anterior margin finely serrate but not medially indented. Anterior commissure broadly sulcate. Costae rounded and crowded, five of the larger ones to a side on the dorsal valve and numerous short intercalations. Interarea strongly apsacline.

Ventral valve evenly and moderately convex in lateral profile; broadly and fairly strongly convex in anterior profile. Sulcus not defined, the median swollen portion occupied by costae. Dorsal valve fairly strongly convex and having a wide but poorly defined median sulcus deepest at the anterior where it forms a broad swale in the anterior half.

Measurements in mm.—Length 7.2; dorsal valve length 6.0; hinge width 6.5; midwidth 8.4; thickness 3.9.

Locality.—P1434.

Types.—Figured specimen: USNM 550741.

Comparison.—This species is larger than *A. crassa,* n. sp., but is more finely costate and does not have the deep median sulci of that species. *Argyrotheca* sp. 1 differs from *A. lutea* in being much larger, more transverse, with less prominent sulci, in having a serrate anterior margin and stronger costation. *Argyrotheca* sp. 1 is most like *A.* sp. 2 in size and dimensions but it does not share the strong development of median sulci and more distant costae of that species.

Argyrotheca sp. 2

Plate 23, figures 20-24

Large for the genus, transversely rectangular in outline with the hinge narrower than the midwidth; sides moderately rounded, anterior margin broadly rounded and finely serrated. Color an even pale yellow. Costae strong and broad, separated by spaces of about equal width to the costae; about 5 primary costae on a side; anterior intercalations irregular. Ventral valve with a median sulcus, poorly defined, shallow but marked medially by a space deeper than the others. Dorsal valve provided with a broad, shallow sulcus widening from the beak, forming a broad sulcation of the anterior commissure, and occupied by four costae, two of which meet the beak.

Measurements in mm.—Length 6.3; dorsal valve length 5.7; hinge width 7.7; midwidth 8.2; thickness 3.6.

Locality.—P630.

Types.—Figured specimen: USNM 550742.

Comparison.—This species is most like *Argyrotheca* sp. 1 in its dimensions and general form and thus is different from *A. lutea* (Dall) and *A. crassa,* n. sp., as explained under those species. It was at first thought that *A.* sp. 1 and sp. 2 were the same but the stronger costation and development of the median sulci in one of them is so different that they could not be united. A suite of specimens of each will be needed to determine the true specific characters.

Argyrotheca sp. 3

Plate 23, figures 1-5

Small, subrectangular in outline with nearly straight sides and well rounded, serrate anterior margin. Pale yellowish white. Ventral valve moderately convex; dorsal valve gently convex. Costae narrowly rounded, distant and separated from each other by spaces wider than the costae. Costae numbering six on a side and an occasional one intercalated. Sulcus on each valve beginning at the beak and widening anteriorly where they are occupied by two short intercalated costae, the largest of which extends to above midvalve.

Measurements in mm.—Length 3.9; dorsal valve length 3.5; hinge width 4.8; midwidth 4.7; and thickness 2.6.

Locality.—GERDA 28-VII—1958.

Types.—Figured specimen: USNM 550740.

Discussion.—In size, shape and color this species is similar to *A. lutea* (Dall) but differs in its serrate anterior margin, the thin and distant costae and smaller size. The costation of *A*. sp. 3 is unlike that of any described Recent species.

A ventral valve identical to that of the above described specimen is in the national collection (USNM 549246). The depth at which the specimen was taken is recorded merely as "deep" and it comes from English Harbor, Antigua. The specimen from Tongue of the Ocean is unusual for the great depth from which it was dredged. Most specimens of *Argyrotheca* in the national collection and those dredged by GERDA and PILLSBURY are from much shallower water.

Argyrotheca sp. 4

Plate 31, figures 1-12

Minute, wider than long, greatest width near midvalve; hinge straight but slightly narrower than the midwidth. Valves unequally convex, the ventral valve having the greater depth. Anterior commissure gently sulcate. Interarea strongly apsacline. Surface multicostate, the costae usually not reaching the beak, fairly broad and low, seven or eight on the ventral valve. Costae opposite as usual in the genus. Foramen large, margined by narrow, elevated deltidial plates. Color yellowish.

Ventral valve marked medially by an obscure fold bearing a medial depression that is occupied by two intercalated costae. Dorsal valve gently sulcate, the sulcus occupied by one or more costae.

Ventral valve interior with wide teeth, a prominent apical plate and long thin median septum reaching nearly to the anterior margin. Anterior end of septum marked by two pits. Dorsal valve with narrow, slitlike sockets; loop reaching valve floor posterior to midvalve, obscure on floor anteriorly and elevated again where it meets the anterior end of the median septum.

Measurements in mm	length	dorsal valve length	mid-width	hinge width	thickness
550759a	2.0	1.6	2.4	2.1	1.3
550759b	2.0	1.8	2.3	1.7	1.3
550759c	2.1	1.7	3.0	3.1	?

Locality.—SILVER BAY Station 961.

Diagnosis.—Minute *Argyrotheca* with sulcate anterior margin and broad irregular costae.

Types.—Figured specimens: USNM 550759a-c.

Discussion.—Although this species is represented by a fair number of specimens and the internal characters suggest maturity, I hesitate to name it. In size it suggests *A. woodwardiana* (Davidson) and *A. schrammi* (Crosse and Fischer). The former is nearly smooth and square in outline, and the latter is ribbed and with a strongly scalloped margin and transversely rectangular. It is like *A. rubrotincta* (Dall), another small species, but that

120

species is larger and with a gaily colored shell of yellow costae with scarlet interspaces.

Argyrotheca spp. indet.

Under this heading are placed specimens not identifiable with known species and which are too few on which to base new species or to make extended descriptions. They are noted here for the sake of completeness.

P630. It is 0.9 mm long by 1.3 mm wide, not costate but strongly bilobate with a median indentation. It could be the young of either *A. barrettiana* or *rubrocostata* with which it was found but there is no way to make sure because the immature of these species are not well known. (USNM 550733.)

P867.—This lot consists of three small, yellowish specimens, the largest of which has the following measurements in mm: length 1.4, width 2. This specimen is not costate but the next to largest specimen is costate with three costae on each side. The third and smallest specimen is bilobate and non-costate. These suggest immature specimens but no species assignment is possible with present knowledge. (USNM 550734.)

P848.—Two specimens both of which are grayish yellow in color comprise this lot, one an adult and the other immature but not appearing to be related to the larger one. The latter is a ventral valve, subquadrate in outline with four costae on a side, the costae becoming obsolete along the anterior margin. The sulcus is fairly wide and occupied by a single costa. The pedicle collar is thick and long but the median septum is low and reaches beyond midvalve. This specimen seems to be unique and is not related to *A. lutea* (Dall), the costae of which continue to the anterior margin. (USNM 550735a.)

The small specimen has four costae on one side and five on the other. The costae are strong, rounded and make a strongly serrated margin. The anterior ends of some of the costae show incipient bifurcation, which suggest that this is not a fully grown specimen. (USNM 550735b.)

P854.—This lot consists of two very small specimens attached to bits of coral. The two specimens belong to the same species. The larger one is rectangular in outline, medially sulcate and incipiently costate. The median sulcate region and lateral incipiently costate parts are bright red, while the remainder of the shell is yellowish. These specimens suggest the young of *A. rubrotincta* (Dall).

Genus *Megathiris* D'Orbigny, 1847

Megathiris detruncata (Gmelin)

Plate 23, figures 8-19

Megathryis d'Orbigny, 1847: 269.—Dall, 1920: 330.—Thomson, 1927: 213.—Elliott *in* Williams, 1965: H831.

This genus is externally similar to *Argyrotheca* in size and costate ornament but may be readily separated from it. *Megathiris* is provided

121

with three or more ridges in the dorsal valve, whereas *Argyrotheca* has only a median septum. *Megathiris* is fairly common in the Mediterranean, but only one report of it in the Caribbean has been made. Dall (1920: 331) reported *M. detruncata* (Gmelin) from Guadeloupe. Dall stated that "The specimen (s) received from Ancey was labeled *Argiope cordata* (a Mediterranean species of *Argyrotheca*) and said to have been collected by Marshall. That it really came from Guadeloupe may well be questioned." Inasmuch as Mediterranean elements such as *Lacazella* live in the Caribbean, occurrence of *Megathiris* there is possible.

Megathiris has not been reported as a fossil from any of the Tertiary formations of the Caribbean or the Gulf Coast of the United States. The genus is an ancient one, as it dates back to the Cretaceous. I record these notes as an encouragement to Caribbean researchers to look for the genus.

Types.—Figured hypotypes: USNM 199368a, b.

Family Platidiidae Thomson, 1927

Genus *Platidia* Costa, 1852

Platidia anomioides (Scacchi and Philippi)

Plate 20, figures 11-19; plate 33, figures 15-17

Orthis anomioides Scacchi and Philippi, 1844: 69, pl. 18, figs. 9a-g.
Terebratula appressa Forbes, 1844: 193.
Platidia anomioides (Scacchi and Philippi) Costa, 1852: 48, pl. 3, fig. 4; pl. 3bis, fig. 6.—Dall, 1870: 14, figs. 20, 21.—Davidson, 1880: 55, pl. 4, figs. 10, 11.—Zittel, 1880: 708.—Deslongchamps, 1884: 160.—Fischer and Oehlert, 1891: 92, pl. 8, figs. 14a-g.—Thomson, 1927: 217; Hertlein and Grant, 1944: 106.
Morrisia anomioides (Scacchi and Philippi) Davidson, 1852: pl. 14, fig. 29.
Platidia (Morrisia) anomioides (Scacchi and Philippi) Davidson, 1870: 405, pl. 21, figs. 15, 15a.
Platydia anomioides (Scacchi and Philippi) Davidson, 1887: 152, pl. 21, figs. 15-19.

Platidia anomioides is fairly large for the genus, nearly circular in outline and with valves of uneven curvature, the ventral valve is always convex but the dorsal valve varies from gently concave to gently convex or posteriorly convex and anteriorly concave. The shape usually depends on the attachment of the shell, whether closely appressed to a rough surface or whether it has had unimpeded freedom to grow. The most important feature of *Platidia* is its pedicle opening. This is generally small in the ventral valve but large in the dorsal valve. This encroachment of the pedicle on the dorsal valve seriously affects the interior of that valve to the extent that the loop has become highly modified. Although the pedicle opening shared by both valves is very large, the pedicle itself is confined to the ventral valve. The pedicle opening is covered by a thick integument.

The peculiar form of the pedicle affects the interior of both valves.

122

The ventral valve has strong teeth. The musculature is unlike that of the majority of articulate brachiopods in having enormous muscles attached to the pedicle. The diductors are small and may attach to the integument over the pedicle opening, which then becomes part of the apparatus used to open the valves. In the dorsal valve a small set of pedicle or adjustor muscles occurs and the usual adductors that divide into two as they pass from the ventral to the dorsal valve are also present. The loop is aborted and consists of descending branches attached to a short, high median septum, sharp curved crural process and two prongs at the end of the septum that are taken to be the remnants of the ascending branches. The lophophore is bilobed and strongly spiculated. The species is rare in the Gulf of Mexico.

This species has been identified from widely separated regions. It is commonest in the Mediterranean region and in the Atlantic along the coasts of Portugal and the northern part of West Africa. It has also been identified in the Indian Ocean although the specific assignment may well be questioned. It is not common in the Caribbean or Gulf regions but is widely distributed. It occurs as a fossil in Tertiary rocks of Cuba.

Localities.—P365, P587, P594, P736, P861, P877, P905, P929; Blake 16; Oregon 4570; Alaminos 9.

Types.—Figured hypotypes: USNM 550528a-c, 550760.

Other localities represented in the national collection are: off Fernandina (Beach) Florida; off Havana, 119 fathoms (= 218 meters) and off Grenada at 291 fathoms (= 533 meters).

Platidia clepsydra Cooper
Plate 33, figures 19-27

Platidia clepsydra Cooper, 1973a: 4, pl., figs. 1-25.

This is a very small species having a white shell, and smooth surface except for concentric lines of growth. It has narrow prongs on the median septum. It is nearly circular, an average specimen having length 2.1 mm, width 2.0 mm and thickness 0.8 mm. The type lot was dredged at Hourglass Station M, latitude 26° 24′ N, longitude 83° 43′ W, about 92 nautical miles due west of Sanibel Island, Florida, at 73 meters. It was also taken west-southwest of Egmont Key, west of Tampa, Florida, at 90 meters (R. Cooper collection). Oregon 35.

G. A. Cooper (1973A) describes the musculature of this small, aberrant *Platidia* in detail indicating the role of the unusually large pedicle muscles and the integument covering the delthyrium in the opening and closing of the valves.

Types. — Holotype: USNM 550748a. Paratypes 550748, b, f, 550749a, 55070d.

Platidia davidsoni (Deslongchamps)
Plate 18, figures 12-22; plate 27, figures 3-6

Morrisia davidsoni Deslongchamps, 1855: 443, pl. 10, figs. 20a-d.

Terebratula (Morrisia) davidsoni (Deslongchamps) Reeve, 1861, pl. 10, fig. 42.
Argiope (Zellania) davidsoni (Deslongchamps) Weinkauff, 1867: 290.
Platidia davidsoni (Deslongchamps) Dall, 1870: 143.—Fischer, 1872: 160, pl. 6, figs. 3-9.—Monterosato, 1879: 306.—Davidson, 1887: 154, pl. 21, figs. 23-27.—Fischer and Oehlert, 1891: 100, pl. 8, figs. 15a-15d.—Cooper, 1973c: 21, pl. 9, figs. 49-52.

Small, subcircular to subquadrate in outline, convexo-to concavo-convex in profile, with the pedicle valve much deeper than the dorsal valve; hinge wide, gently curved, narrower than the maximum width which is at midvalve; interareas narrow and short. Pedicle opening shared by both valves; no deltidial plates. Surface marked by crowded pustules of two sizes, those of the dorsal valve somewhat smaller and less numerous than those of the ventral valve.

Ventral valve swollen medially, the swelling diminishing in all directions from the middle; beak small and obtusely pointed. Anterior and lateral profiles forming a high dome; anterior slope moderately steep and the lateral slopes shorter and steeper.

Dorsal valve variable from gently convex to gently concave, often irregularly undulated from contact with the substrate; foramen large, irregularly semicircular and occupying about 1/5 the valve length. Umbonal region slightly swollen.

Ventral valve interior with nub-like teeth located just anterior to the distal ends of the interareas; median ridge short and thick; muscle region slightly thickened.

Dorsal valve with long teeth at the angles of the foramen; hinge plates long, concave; crural processes short and blunt; loop thin and delicate, attached to the median septum; in length measuring 1/2 the valve length and anterior processes or points thin.

Measurements in mm	length	dorsal valve length	width	thickness
550544a	5.2	4.9	5.3	ca. 1.5
550545a	4.3	3.8	4.9	2.0

Localities.—G190, G304, GERDA-VII-1958; OREGON 4570.

Types.—Hypotypes: USNM 550544a, 550545a. 550762a, b.

Diagnosis.—Pustulose *Platidia.*

Discussion.—This species differs from *P. anomioides* (Scacchi and Philippi) in being strongly pustulose on the ventral valve whereas that of the other species is marked by concentric lines only. Furthermore, the loop of *P. davidsoni* is supported by stouter hinge plates, but the crural processes and anterior forked process are more delicate than those of *P. anomioides.*

The occurrence of this Mediterranean and eastern Atlantic species in the Caribbean and Gulf of Mexico is more evidence of the close connection between the two areas.

From OREGON Station 4570, two specimens of the new genus *Ecno-*

miosia bear attached *Platidia;* one is the smooth *Platidia anomioides* and the other is *P. davidsoni* covered by pustules.

The Caribbean and Gulf of Mexico specimens of *P. davidsoni* are somewhat more rounded than those figured from the Mediterranean. It occurs also off the Argentina coast, southeast of Mar del Plata (Cooper, 1973c).

Family Kraussinidae Dall, 1870

Genus *Megerlia* W. King, 1850

Megerlia echinata (Fischer and Oehlert)

Plate 17, figures 14-22

Mühlfeldtia echinata Fischer and Oehlert, 1890: 73; 1891, p. 90, pl. 7, figs. 13a-g.
Pantellaria echinata (Fischer and Oehlert) Dall, 1920: 336.
Megerlia echinata (Fischer and Oehlert) Atkins, 1961: 89-94.

This is a rare species in the Caribbean region but is better known from off the northwest coast of Africa. As identified at present it has a wide distribution: southwest of the Cape of Good Hope, Africa; off New South Wales, Australia; off Barbados; and off Sand Key, Straits of Florida. The PILLSBURY has added another location for this unusual brachiopod: northeast of Caracas, Venezuela.

A single specimen (USNM 550527) taken by the PILLSBURY is unusually well formed. Although the dorsal valve is abraded, it is not greatly distorted and is moderately convex. Inside the ventral valve is a small anteriorly excavated plate supported by a short median septum. The teeth are small and a narrow interarea forms the margin of the wide pedicle opening. Inside the dorsal valve the loop is well preserved and consists of short crura bearing sharp crural processes. The short descending lamellae join a broad projection that extends anterolaterally and toward the ventral valve. These two extensions are united dorsally by a narrow ribbon and the two projections are attached to the anterior end of the strong median septum. The socket ridges are unusual because each bears a small posteriorly directed projection or denticle that articulates with sockets formed by the outside of the apical plate and the inside shell wall of the ventral valve. The valves are thus articulated by teeth and sockets in both valves, the teeth of the dorsal valve being the stronger and probably the more effective of the two.

Locality.—P739.

Types.—Figured hypotype: USNM 550527.

Discussion.—The muscle system of a closely related species is illustrated on figure 8. This is *Megerlia truncata* (Linnaeus) and the specimen is a ventral valve (USNM 174943a). This shows the great development of the adjustor muscles at the expense of the diductors, which are normally the largest muscles in the ventral valve. These may be compared with the similar arrangement of *Argyrotheca rubrocostata* illustrated on the same figure.

125

Family Macandreviidae Cooper, 1973c

Aberrant Terebratellacea having the hinge plates extended directly to the valve floor and without a median septum in the adult. Loop development as in Dallinidae.

This name was proposed because of the great difference in the cardinalia of *Macandrevia* from those of other Dallinidae. The cardinalia of *Dallina* have hinge plates united with a median septum in the adult. *Terebratalia* also has aberrant cardinalia, quite different from those of *Dallina* and *Macandrevia*.

Genus *Macandrevia* W. King, 1859

Macandrevia W. King, 1859: 261.—Thomson, 1927: 239.—Muir-Wood, Elliott and Hatai *in* Williams, 1965: H837.

This genus is usually easy to recognize because of several exterior and interior characters. Its foramen is usually strongly excavated and the deltidial plates at best are vestigial. It is usually rectimarginate and without ornament other than concentric lines of growth. Inside the ventral valve the teeth are supported by prominent dental plates. In the opposite valve the loop is dalliniform, the outer hinge plates are reduced and the crura are supported by plates that extend directly to the valve floor and bound a narrow notothyrial chamber.

Macandrevia has a wide but peculiar geographic distribution. It occurs in the North Atlantic from Greenland to the Spanish Sahara. Its range along the West African coast has been extended to near Sao Tome just west of Gabon by the R/V PILLSBURY (station P314). It is rare in the Mediterranean but was a member of the Mediterranean fauna in the Tertiary. It occurs on the east coast of North America from Georges Bank off Massachusetts to New Jersey but has not been reported farther south. Although not yet seen in the Caribbean region it should be expected there because it occurs in the Gulf of Panama and southward along the west coast of South America to the Antarctic. It also occurs northward to San Diego off the west coast of the United States. This genus occurs at several places around the Antarctic continent. Except for the occurrences along the west coast of North and South America, it is not now known in the Pacific.

Fossil *Macandrevia* is known in Mediterranean Tertiary deposits but also has been taken from the Tertiary of Japan (Hatai, 1940: 266-269). It has not been seen in the Tertiary of the United States Gulf Coast or that of the Caribbean.

Macandrevia occurs in shallow water as well as at abyssal depths. It is reported by Dall in one fathom (= 1.83 meters) of water at Bodo, Norway. Along the Pacific coast of the Americas it ranges from 122 fathoms (223 meters) to 2222 fathoms (4066 meters). *Macandrevia novangliae* appears in the area covered by this paper and ranges from 1837 to 2492 meters.

The greatest abundance of this genus is in the colder waters of the North Atlantic and around Antarctica. In the intermediate occurrences it

126

is usually in deep water. *Macandrevia* should ultimately be found in deep Caribbean waters because the present Caribbean brachiopod fauna appears to have been inherited from the Tertiary with Tethyan affinities.

Macandrevia novangliae Dall

Plate 26, figures 1-11

Macandrevia cranium, new var. *novangliae* Dall, 1920: 355, 356.

Small for the genus, pale to dark brown in color; pentagonal in outline; lenticular in profile but with the ventral valve deeper than the dorsal valve. Sides gently rounded; anterior margin rounded in young to truncate in old specimens; apical angle 88°-100°. Anterior commissure rectimarginate. Foramen opening into the delthyrium which is only slightly restricted by rudimentary deltidial plates. Surface marked by concentric lines of growth.

Ventral valve moderately convex with maximum convexity in the posterior half; anterior profile broadly convex, the sides sloping gently nearly to the margins where they are very steep. Beak small, erect, only moderately produced posteriorly; umbo evenly swollen. Anterior slope gentle.

Dorsal valve gently and evenly convex in lateral profile but with a flattened top in anterior profile; sides short and steep. Umbonal and median regions gently swollen; anterior slope short and steep.

Ventral valve interior with small teeth buttressed by thick dental plates. Muscles not strongly impressed. Dorsal valve with thin socket ridges, poorly defined outer hinge plates and supporting plates fairly widely separated. Pallial trunks well developed.

Measurements in mm	length	dorsal valve length	width	thickness	apical angle
78069	15.0	13.1	13.2	8.5	93°
78340	12.8	11.5	10.8	6.8	90°
44968	14.2	12.3	11.7	8.0	88°
550625	11.5	10.2	11.3	5.7	100°

Localities.—ALBATROSS 2035, 2208, 2220, 2535, 2682, 2706.

Diagnosis.—Small, roundly pentagonal, with fairly stout shells.

Types.—Holotype USNM: 78069. Figured hypotypes: USNM 49068a, b, 550625.

Comparison.—This species, which was suggested as a variety of *Macandrevia cranium* (Müller) by Dall, was never formally described and was not illustrated. It need be compared only to the smaller species of the genus such as *M. tenera* (Jeffreys) from the North Atlantic. It is smaller than *M. cranium* (Müller), to which Dall had referred it. From *M. tenera* it differs in shape as it is rather more chunky and strongly pentagonal than that species. It is stouter, wider and rounder than *M. tenera,* which is somewhat elongated and thin-shelled. *Macandrevia novangliae* in exterior form is perhaps closest to *M. euthyra* (Philippi), a Pliocene species from Sicily. The fossil species, however, is usually definitely larger.

Family Dallinidae Beecher, 1893

Genus *Dallina* Beecher, 1893

Dallina floridana (Pourtalès)

Plate 4, figures 1-4; plate 11, figures 20-27; plate 29, figures 14-24

Waldheimia floridana Pourtalès, 1868: 127.—Dall, 1871: 12, pl. 1, fig. 3; pl. 2, figs. 1-3.—Davidson, 1887: 59, pl. 12, figs. 1-5.
Dallina floridana (Pourtalès) Beecher, 1893: 382, pl. 1, fig. 45.—Dall 1920: 359.

This is perhaps the easiest recognized brachiopod in the Gulf of Mexico and Caribbean regions because of its strongly triangular outline, pale yellow color, and wide anterior which has a fold within the sulcus. The adult ventral valve is without dental plates and the dorsal valve has a strong median septum, and concave inner hinge plates that unite with the septum. The descending branches of the loop are very thin and delicate but they support a broad, hood-like transverse ribbon. This species commonly occurs with *Tichosina cubensis* (Pourtalès) and their dissociated valves are often mingled in the bottom debris.

Localities.—G241, G242, G246, G482, G579, G835, G838, G839, G863, G864, G942, G972, G973, G974, G1029, G1036, G1102, G1125, G1312, G1314; P587, P988; OREGON 2644, 2646, 4574, 4994, OREGON II Station 11133; SILVER BAY 2416, 2418, 2426, 2427, 3494, 3513, 5181; ALBATROSS 2388. West-southwest of Dry Tortugas at 457 meters (R. Cooper collection).

Types.—Figured hypotypes: USNM 110861b, 432772, 550582, 550583, 550616, 550758a, b.

Discussion.—The loop development of *Dallina* is quite complicated and when young stages are found they are difficult to identify. The young *Dallina* develops a median pillar in the dorsal valve on which a ring develops (see plate 3, fig. 24). Continued growth enlarges the ring at the same time the descending branches join the pillar. While the ring is enlarging, part of the septum is resorbed to free the now joined ring and descending branches to form the loop. The original ring becomes the ascending and transverse bands of the loop. An intermediate genus *Fallax* has been recognized, and is an adult dallinid with loop stage arrested at the time of late attachment to the median septum.

Dallina floridana is often found with *Tichosina cubensis* (Pourtalès) in the Straits of Florida and in occurrences in the Caribbean. *Dallina* is common in the Straits of Florida but also occurs in the Caribbean and Gulf of Mexico where it is rare.

Family **Ecnomiosidae** Cooper, new family

Roundly oval, biconvex, with large foramen and disjunct deltidial plate; cardinalia having inner hinge plates attached to the short median septum;

adult loop long, anteriorly free but with ascending branches attached to the median septum by a ring formed by the transverse band; descending branches of loop attached to the median septum in juvenile and early adult stages of loop development.

Ecnomiosa Cooper, new genus

Shell slightly above an inch and a half in length (39 mm), pentagonal to subcircular in outline with strongly rounded sides and somewhat nasute anterior margin; beak short, truncated by a large foramen, submesothyridid in position; deltidial plates rudimentary, disjunct. Anterior commissure rectimarginate to faintly sulcate. Color of shell white where abraded but covered with a rich brown periostracum. Surface marked by closely spaced concentric lines. Punctae crowded, small.

Ventral valve interior with small stout teeth and strong but short dental plates. Muscle field small, anterior to delthyrial cavity, diductor scars moderately deeply impressed; adductors attached on a short low median ridge. Vascula media narrow extending directly anteriorly.

Dorsal valve with confined cardinalia, the socket ridges low but strong; sockets small; outer hinge plates narrow; thickened; crural bases moderately wide; inner hinge plates flatly concave and attached to a very short median septum that reaches its crest about 1/6 the valve length from the beak; septum with a steep anterior face sloping anteriorly and extending along the valve floor to end about 1/3 the valve length from the beak. Crura very short, round in section and giving off flattened, sharp crural processes; descending ribbons slender, spiny at their distal extremities; ascending branches widening posteriorly, with remnantal hood, branching into two at their crest, the two branches diverging and extending dorsally, then uniting to form a short rounded pillar that joins with the crest of the median septum. Medium-size specimens having descending branches of loop attached to septum as in *Terebratalia*. In the adult these are resorbed.

Type-species.—*Ecnomiosa gerda* Cooper, new species.

Derivation.—Greek *eknomios* = unusual, marvelous.

Diagnosis.—Medium sized to large brachiopods; adults, having a short median septum to which is attached a ring formed by division of the hood into two prongs that join dorsally to form a short pillar uniting with the crest of the median septum.

Comparison and discussion.—The loop of this brachiopod is so unusual as to set the genus apart from all others known. The exterior, however, suggests a *Terebratalia* in the large and much resorbed foraminal margin and the tendency toward sulcation of the anterior commissure. The shell surface is concentrically marked, not radially as usual in most species of *Terebratalia* except *T. coreanica*. The cardinalia are unlike those of *Terebratalia* in which there are no inner hinge plates and no median septum in the adult. The cardinalia of *Ecnomiosa* suggest those of *Dallina* or *Terebratella* but are not as concave as these and the median septum is much

129

shorter than in either of those genera. The entire dorsal valve in its cardinalia and loop is unique.

Four specimens of this genus were taken by the GERDA two complete ones and two damaged ones. A loop was developed from one of the good specimens and one of the broken ones yielded good cardinalia, muscle scars and pallial impressions (see plate 34). Additional specimens came in the Bullis collection. The genus was taken at three locations. At none of them was it abundant but one lot included a few young specimens and one immature individual with loop. These specimens gave much detail on the anatomy and an insight to the development of the loop. Nevertheless, complete understanding of the genus will only come from a more extensive growth series.

The youngest specimen (USNM 550589b) measures 8.4 mm in length by 8.2 mm in width. The beak of the ventral valve narrows rapidly and the foramen is completely open without a trace of deltidial plates. In the dorsal valve of this small specimen the hinge plates are well differentiated. The socket ridges are fairly strong and the outer hinge plates are narrow but thickened. The crural base is a well defined low ridge that extends to the apex; the inner hinge plates are large, deeply concave and join the low thick median septum. The crura are short, thick but flattened. The crural processes are bluntly pointed. The descending lamellae are broad, moderately curved and unite anteriorly. On their inner edge occurs an elongate hood, narrowly cleft medially for half its length on the dorsal side. Its free edge along this cleft bears fairly long spines decreasing in size posteriorly. The hood is conical, flattened on its posteroventral surface and with a narrow triangular opening at its posterior end. The dorsad side of the cone narrows and the sides unite posteriorly and join the median septum. The loop of this specimen is in the campagiform stage. This strongly resembles the campagiform stage of *Dallina* and *Macandrevia,* but is a larger shell at this stage.

The specimen next in size (USNM 550590a) measures 17.6 mm in length and 17.1 mm in width, twice the dimensions of the preceding one and with a loop almost in adult condition. In this specimen the hinge plates have attained their adult condition, solidly fused, thick, not clearly differentiated, and the whole forming a nearly horizontal platform. The loop, on the other hand although nearly adult, retains some juvenile features. The descending branches are slender ribbons and are attached to the median septum by lateral branches. Were it not for the peculiar development of the hood with its attachments, the loop would be classified as in the terebrataliform stage. The ascending branches are broader than the descending ones and join a flattened plate, a remnant of the hood. Distally this sends two prongs dorsally that form a ring, the distal side of which attaches to a stout pillar built up from the crest of the median septum. At this point on the septum are joined the attachments from the ascending elements of the loop and the lateral bars that bind the descending processes to the median septum. In this specimen one of the lateral bars from the septum to one of the descending branches is partially resorbed.

130

A specimen (USNM 550600) measuring 30.7 mm in length and 28.6 mm in width has a juvenile loop. In this specimen, although fully adult in size, the lateral branches attaching the descending lamellae to the median septum are thin but well developed. The holotype, which is only two millimeters longer than the preceding, has small projections on the descending lamellae, remnants of the nearly completely resorbed lateral branches. Two small projections also appear on the septum that are remnants of the lateral branches. It is likely that descendants of this brachiopod will ultimately resorb all connections of the loop to the septum and might resorb the ring to produce a completely free hanging loop like that of adult *Dallina*.

Ecnomiosa gerda Cooper, new species

Plate 34, figures 1-15; plate 35, figures 1-14

A medium to large brachiopod, pentagonal in outline in old specimens but subcircular in the young. Sides rounded; anteriorly narrowly rounded to subnasute; apical angle slightly greater than a right angle. Beak blunt, narrowly rounded and with a large submesothyridid foramen; deltidial plates small, rudimentary. Valves unequally convex, the ventral valve having the greater depth. Surface marked by closely crowded strong concentric lines. Punctae 144 per square millimeter.

Ventral valve fairly strongly convex in lateral profile with the maximum convexity just posterior to midvalve; anterior profile a strongly convex dome with long steep sides. Umbonal regions strongly inflated, the swelling extending over the whole valve; anterior slope steep. Posterolateral slopes narrowly curved.

Dorsal valve moderately convex in lateral profile with the maximum convexity at about midvalve; anterior profile broadly and gently convex, with the median region slightly flattened in old individuals; entire valve moderately inflated but the median region depressed slightly to form a shallow sulcus in a very old specimen. Flanks swollen and steep.

Interior as described for the genus.

Measurements in mm	length	dorsal valve length	maximum width	thickness	apical angle
550510a	32.2	27.9	27.6	?	93°
550510b	27.0	23.3	26.0	16.0	91°
550510c	22.4	19.4	22.5	11.1	96°
550510d	?	28.9	31.8	?	?
550606a	39.5	34.9	37.8	20.8	105°
550606b	35.0	31.9	33.5	18.6	102°

Locality.—GERDA 114; OREGON 4570; OREGON II 10962.

Diagnosis.—*Ecnomiosa* elongating with age and marked by closely crowded concentric lines.

Types.—Holotype: USNM 550510a. Figured paratypes: USNM 550510b, c,

550589a, 550590a, b, 550600, 550606a. Unfigured paratypes: USNM 550589b, 550606b.

Comparison.—No other species of this genus is known to which this one may be compared. Named for the R/V GERDA.

Suborder THECIDEIDINA Elliott, 1965

Superfamily THECIDEACEA Gray, 1840

Family Thecidellinidae Elliott, 1958

Genus *Thecidellina* Thomson, 1915

Thecidellina barretti (Davidson)

Plate 3, figures 19-21, 22-26

Thecidium barretti Davidson, 1864: 17, pl. 2, figs. 1-3.—Crosse and Fischer, 1866: 272.—Dall, 1870: 151; 1871: 42.—Davidson, 1887: 162, pl. 23, figs. 9-11. *Thecidellina barretti* (Davidson), Thomson, 1915: 462.—Dall 1920: 283.— Thomson, 1927: 139.

This is fairly common in the Caribbean region but may easily be overlooked because it is small and cemented on coral or other solid substance. *Thecidellina* and *Lacazella,* which is still rarer in the Caribbean, are the only two known modern cemented articulate brachiopods.

Thecidellina is usually white, small, subpentagonal or triangular in outline and with a long beak having a flat interarea. Cementation is by the beak and part of the ventral valve. Internally the ventral valve has two plates, the hemispondylium defining part of the muscle region. The dorsal valve is marked by a strong median ridge, widest anteriorly and tapering posteriorly. The lophophore is primitive and defined as schizolophous.

Localities.—P630; NGS-MRF-33.

Types.—Figured hypotypes: USNM 550540a, b, 550541.

Discussion.—*Thecidellina* in the Caribbean region has an ancient history. It is known from the Eocene of Cuba and has been taken from the Miocene as well. It is also known from the Eocene of the Gulf Coast and South Atlantic coast of the United States. It has recently been reported from the Eocene of Pyrenees in France by Pajaud and Tambareau (1970). It is also known from the Eocene and Miocene of the Pacific (Cooper 1971).

Family Thecideidae Gray, 1840

Genus *Lacazella* Munier-Chalmas 1881

Lacazella caribbeanensis Cooper, new species

Plate 4, figures 12-19

Thecidium mediterraneum Davidson, 1864: 21, pl. 2, fig. 5; 1886: 158. *Lacazella mediterranea* (Risso) Meile and Pajaud, 1971:470, pl. 1, figs. 2-4 (not fig. 1:=*Thecidellina?*).

Small, transversely oval in outline with the maximum width at about midvalve; sides somewhat narrowly rounded; anterior margin gently rounded to nearly straight; hinge narrower (about half) than the valve width, beak somewhat narrowly pointed; pseudodeltidium (?) narrow, not strongly convex; shell substance punctate.

Ventral valve moderately convex and with a large area of attachment; anterior half somewhat flattened and moderately geniculated. Dorsal valve moderately swollen medially and with the sides and anterior somewhat flattened.

Ventral valve interior with strongly papillose inner rim but most of the valve floor smooth to obscurely radiately striated; teeth small; hemispondylium narrow, not complete anteriorly and consisting of two separate thin plates bounding a narrow chamber with strong median septum not reaching the anterior margin of the structure; hemispondylial plates pointed distally.

Dorsal valve interior with flattened and strongly papillose rim; ascending apparatus narrow and deep anteriorly and with smooth margins; descending apparatus (interbrachial lobe) forming a wide and smooth margined crescent; median ridge of the ascending apparatus not grooved and not spinose; median elements of the descending apparatus short and bluntly pointed; bridge broad and with a fairly deep and wide marsupial notch. Cardinal process (cardinal extension of Elliott 1948) rather narrow and fairly long.

Measurements in mm	length	dorsal valve length	width	thickness
549449a	3.0?	2.1	2.7	ca 2.0

Locality.—Smithsonian-Johnson Exped. Station 52, off Rio Bueno, north central Jamaica.

Diagnosis.—Small *Lacazella* with hemispondylium attached to the valve floor and the ascending apparatus narrow anteriorly and with smooth margins.

Types. Holotype: USNM 549449a; figured paratype USNM 549448a, 59449b.

Discussion.—Although Davidson (1887: 158) mentioned this species as having been taken by the BLAKE from the Gulf of Mexico, he was actually referring to a specimen of *Thecidellina.* He reported *Lacazella* from the Caribbean region off Jamaica, and Jackson, Goreau and Hartman (1971) reported it as common on the north side of Jamaica. The specimens here described extend its range to the Dominican Republic. The genus is a holdover from the Tertiary when it was fairly common in Caribbean and Atlantic waters. It is known from the Eocene to Miocene of Cuba (undescribed) and from the Eocene of the Gulf and Atlantic coasts of the United States.

The new species is smaller than *Lacazella mediterranea* (Risso) (plate 4, figures 5-11), the commonest modern species, but in shape and other characters it would be difficult to separate the two. When the interior is investigated the differences are striking. In the ventral valve of *L. mediterranea*, the hemispondylium is generally a bilobed plate with strong median septum, but in *L. caribbeanensis* it consists of two plates, not connected, bounding a narrow chamber divided by a sharp, low median septum. The sides of the hemispondylium are bounded by plates distally, sharply pointed. In the dorsal valve the differences are equally striking. The ascending elements have smooth margins and the median ridge between them is not grooved but is narrow and rather septate than tongue-shaped as it is in *L. mediterranea*. Other elements of the cardinalia are also smooth or nearly so and the descending processes at midvalve are short and bluntly pointed rather than long as in the Mediterranean species. The interior of both valves of *L. mediterranea* are strongly papillose or ridged whereas those of *L. caribbeanensis* have very subdued markings.

The only other species of *Lacazella* now known is *L. mauritiana* Dall from the Island of Mauritius in the Indian Ocean. Internally this species is very similar to *L. mediterranea*. It is thus distinguished from the Caribbean species by the same characters: larger size, strongly granulose interior and very wide ascending element anteriorly and serrate or spiny margins to both ascending and descending elements.

REFERENCES

ATKINS, G.
1961.   The generic position of the brachiopod *Megerlia echinata* (Fischer and Oehlert). J. mar. biol. Assoc. U.K., *41*:89-94, t. figs.

BEECHER, C. E.
1983.   The development of *Terebratalia obsoleta*. Trans. Conn. Acad. Arts Sci., *9*:392-399, pls. 2, 3.

BRODERIP, W. J.
1835.   Descriptions of some new species of Cuvier's family of Brachiopoda. Trans., Zool. Soc. Lond. *1*:141-144, pls. 22, 23.

BROMLEY R. G. AND F. SURLYK
1973.   Borings produced by brachiopod pedicles, fossil and Recent. Lethaia, *6*(4): 349-365.

BROOKS, W. K.
1879.   The development of *Lingula* and the systematic position of the Brachiopoda. Chesapeake Zool. Lab. Sci. Res., *1*:35-112, 6 pls.

COOPER, G. A.
1934.   New Brachiopods. Reports on the collections obtained by the first Johnson-Smithsonian deep-sea expedition to the Puerto Rican Deep. Smithson. Misc. Coll., *91*(10): 5 p, 1 pl.

1954.   Gulf of Mexico, its origin, waters and marine life. Brachiopoda occurring in the Gulf of Mexico. Fish Wildl. Serv., Fish. Bull., *55*:363-365.

1955.   New Brachiopods from Cuba. J. Paleo., 29(1):64-70, pl. 15.

1959.   Genera of Tertiary and Recent rhynchonelloid Brachiopods. Smithson. Misc. Coll., *139*(5):90 p., 22 pls.

1970.   Generic characters of brachiopods. Symposium, North American Paleontological Convention, 1969: The Genus: A basic concept in Paleontology; 194-263.

1971.   Eocene brachiopods from Eua, Tonga. U. S. Geol. Surv. Prof. Paper 640-F; F1-F9, 1 pl.

1972.   Homeomorphy in Recent Deep-sea brachiopods. Smithsonian Contrib. Paleobiol., *11*: 16 p., 4 pls., 5 figs.

1973a.   Brachiopods (Recent). Mem. Hourglass Cruises, *3*(3), 17 p., one plate, 6 figs.

1973b.   Fossil and Recent Cancellothyridacea (Brachiopoda). Sci. Repts. Tohoku Univ., Sendai, Japan, Second Ser. (Geology), Spec. vol. (Hatai Mem. vol.) 6:371-390, pls. 42-46.

1973c.   Vema's brachiopods (Recent). Smithson. Contrib. Paleobiol., *17*:51 p., 9 pls., 5 figs.

COSTA, O. G.
1852.   Fauna del Regno di Napoli ossia enumerazione di tutti gli Animali - contenente la descrizione de nuovi o poco essattemente consciuti - di O. G. Costa (continuata da A. Costa). 19 sections, in 11 volumes. 4° Napoli (1829-) 1832-70 (-86). Classe V, Brachiopoda, 60 p, 9 pls.

135

CROSSE, H.
  1865.   Description d'especes nouvelles de Guadeloupe. *Terebratulina cailleti, Murex abyssicola, Fusus Schrammi, Pleurotoma Jelskii, P. antillarum, Astralium Guadeloupense.* J. Conchyl., ser. 3, *5*:27-38.

CROSSE, H. AND P. FISCHER
  1866.   Note sur la distribution geographique des Brachiopodes aux Antilles. J. Conchyl., 3rd ser. *6*: 265-273.

DALL, W. H.
  1870.   A revision of the Terebratulidae and Lingulidae, with remarks on and description of some recent forms. Amer. J. Conch., *6*(2): 88-168, pls. 6-8, figs. 1-38.

  1871.   Report of the Straits of Florida on the Brachiopoda obtained by the United States Coast Survey expedition, in charge of L. F. de Pourtales, with a revision of the Craniidae and Discinidae. Bull. Mus. Comp. Zool. Harvard, *3* (1): 45 p, 2 pls.

  1873.   Catalogue of the Recent species of the Class Brachiopoda. Proc. Acad. Nat. Sci. Philad., 1873; 177-204.

  1882.   American work on recent Mollusca in 1881. Amer. Nat., *16*(11): 874-887.

  1886.   Report on the results of dredging . . . in the Gulf of Mexico (1877-78) and in the Caribbean Sea (1879-80). XXIX. Report on the Mollusca. Part 1. Brachiopoda and Pelecypoda. Bull. Mus. Comp. Zool. Harvard, *12*: 171-318, pls. 1-9.

  1903.   Contributions to the Tertiary fauna of Florida. Class Brachiopoda. Trans., Wagner Free Inst. Sci., *3*(6): 1533-1540, pl. 58.

  1908.   Reports on the Mollusca and Brachiopoda ["Albatross" Dredging operations in the western Pacific]. Bull. Mus. Comp. Zool., Harvard, *43*(6): 205-487, pls. 1-19.

  1911.   A new brachiopod from Bermuda. Nautilus, *25*: 86, 87.

  1919.   New shells from the northwest coast. Proc. Biol. Soc. Washington, *32*: 249-252.

  1920.   Annotated list of the Recent Brachiopoda in the collection of the United States National Museum, with descriptions of 33 new forms. Proc. U. S. Nat. Mus., *57*: 261-377.

DAVIDSON, T.
  1852.   Description of a few new recent species of Brachiopoda. Proc. Zool. Soc. Lond., *20*: 75-83.

  1855.   A few remarks on the Brachiopoda. Ann. Mag. Nat. Hist. ser. 2, *16*: 425-429, pl. 10.

  1864.   On the Recent and Tertiary species of the genus *Thecidium*. Geol. Mag., *1*: 12-23, pls. 1, 2.

  1866.   Notes on some Recent Brachiopoda dredged by the late Lucas Barrett off the northeast coast of Jamaica, and now forming part of the collection of R. MacAndrew. Proc. Zool. Soc. Lond.: 102-104, pl. 12.

  1870.   On Italian Tertiary Brachiopoda. Geol. Mag., *7*: 359-370, 399-408, 460-466, pls. 17-21.

  1878.   Preliminary report on the Brachiopoda dredged by H. M. S. "Challenger." Proc. Roy. Soc., 27 (188): 428-439.

1880.   Report on the Brachiopoda dredged by H.M.S. "Challenger" during the years 1873-1878. Rep. Sci. Res, Challenger, Zool. *1*:1-67, 4 pls.
1886-   A monograph of Recent Brachiopoda. Trans. Linnean Soc. London,
1888.   parts 1-3, 2nd. Ser. Zool., *4*: 1-248, 30 pls.

DESLONGCHAMPS E. EUDES
1855.   On a new species of *Morrisia, in* Davidson, 1855: 443, pl. 10, figs. 20, 20 a-d (see above).

1884.   Notes sur les modifications a apporter à la classification des Terebratulidae. Bull. Soc. Linn. Normandie, *8*: 77-388, pls. XIII-XXVIII.

ELLIOTT, G. F.
1948.   Palingenesis in *Thecidea* (Brachiopoda). Ann. Mag. Nat. Hist., Ser. 12, *1*: 1-30, 2 pls.

1958.   Classification of .Thecidean brachiopods. J. Paleont., *32*(2): 373.

1965.   Suborder Thecideidina, *in* Williams, *et al*: H858-H862, figs. 742-745 (see below).

FISCHER, P.
1872.   Brachiopodes des cotes océanique de France. J. Conchyl., 3 ser., *2a*: 160-164, pl. 6, figs. 3-9.

FISCHER, P. AND D. P. OEHLERT
1890.   Diagnoses de nouveaux Brachiopodes. J. Conchyl., *38*: 70-74.

1891.   Expeditions scientifiques du Travailleur et du Talisman pendant les années 1880, 1881, 1882, 1883. Brachiopodes. Paris. 140 pp., 8 pls.

FORBES, E.
1844.   Report on the Mollusca and Radiata of the Aegean Sea, and on their distribution, considered as bearing on Geology. Rept. Brit. Assoc. 1843: 130-193.

FOSTER, M. W.
1968.   Harvard University's brachiopod studies on *Eltanin* Cruise 32. Antarctic J. U. S., Sept-Oct. 1968.

GRAY, J. E.
1840.   Synopsis of the contents of the British Museum: 42nd edit. 370 p. (London).

1848.   On the arrangement of the Brachiopoda. Ann. Mag. Nat. Hist. ser 2, *2*: 435-440.

HARDING, J. L. AND W. D. NOWLIN
1967.   Gulf of Mexico, *in* Encyclopedia of Oceanography. Encyclopedia of Earth Sciences, vol. 1. (R. W. Fairbridge editor):i-xiii plus 1021 p. Gulf of Mexico:324-330. Reinhold Publishing Company, New York, New York.

HATAI, K.
1940.   Cenozoic Brachiopoda from Japan. Tohoku Imper. Univ., Sci. Repts., Second Ser. (Geology), *20*: 1-413, 12 pls.

HELMCKE, J. G.
1940.   Die Brachiopoden der Deutschen Tiefsee-Expedition. Wiss. Ergebn. Deutsch. Tiefsee Exped. 1898-99, *24*(3): 215-316, figs. 1-43.

HERTLEIN L. G. AND U. S. GRANT IV
1944.   The Cenozoic Brachiopoda of western North America. Publ. Univ.
Calif. Los Angeles, Math., Phys. Sci., *3*: 1-236, pls. 1-21.

HUXLEY, T. H.
1869.   An introduction to the classification of animals. 147 p., 47 figs. Church-
ill and Sons (London).

JACKSON, J. B. C., T. F., GOREAU, AND W. D. HARTMAN
1971.   Recent Brachiopod-Coralline Sponge Communities and their Paleoe-
cological Significance. Science, *173*(3997): 623-625, 2 figs.

JEFFREYS, G.
1869.   The deep-sea dredgings. Explorations on H.M.S. "Porcupine."
Nature, *1*: 136.

1876.   On some new and remarkable North-Atlantic Brachiopoda. Ann.
Mag. Nat. Hist., ser. 4, *18*: 250-253.

1878.   On the Mollusca procured during the "Lightning" and "Porcupine" Expeditions, 1868-70, Pt. 1. Proc.: Zool. Soc. Lond. 393-416,
pls. 22, 23.

KING W.
1859.   On *Gwynia, Dielasma* and *Macandrevia,* three new genera of Pallio-
branchiata Mollusca, one of which has been dredged in Strangford
Lough. Proc. Zool. Bot. Assoc., Dublin Univ., *1*(3): 256-262.

1868.   On some palliobranchiate shells from the Irish Atlantic. Proc. Dublin Nat. Hist. Soc., *5*: 170-173.

KUHN, O.
1949.   Lehrbuch der Paläozoologie. E. Schweizerbart, Stuttgart. 326pp.,
244 figs.

KÜSTER, H. C.
1843.   Conchylien-Cabinet von Martini und Chemnitz., VII, Brachiopoda, p. 15, pl. 1, figs. 10, 11.

LOGAN, A.
1974.   Life habits of the Recent articulate brachiopod *Argyrotheca bermudana* from Bermuda. Geol. Soc. Amer., Abst. with Programs, *5* (1):
48-49. (To appear as Contrib. No. 519, Bermuda Biol. Sta. for Research).

MEILE, B. AND M. D. PAJAUD
1971.   Présence de brachiopodes dans le Grand Banc des Bahamas. C. R. Acad.
Sci. Paris,   D, *273* :469-472.

MENCKE, C. T.
1828.   Synopsis methodica molluscorum generum omnium et specierum
earum quae in Museo Menkeano adservantur, 91 p. (Pyrmonti).

MONTEROSATO, MARQUIS T. A. DI
1879.   Note sur les espèces du genre *Platidia*. J. Conchyl., 3 ser., *19*: 306.

MORSE, E. A.
1902.   Observations on living Brachiopoda. Mem. Boston Soc. Nat. Hist., *5*:
313-386, pls. 39-61.

MUIR-WOOD, H. M.
1955.   A history of the classification of the phylum Brachiopoda. British
Museum (Nat. Hist.), London, 124 pp., 12 figs.

138

1959.   Report on the Brachiopoda of the John Murray Expedition. Sci. Rept., John Murray Expedition 1933-34, *10*(6): 283-317, 5 pls.

1960.   Homoeomorphy in Recent Brachiopoda: *Abyssothyris* and *Neorhynchia*. Ann. Mag. Nat. Hist., Ser. 13, *3*: 521-527, pl. VII.

1965.   Mesozoic and Cenozoic Terebratulidina *in* Williams, *et al.*: H 762-H 816, figs. 622-695 (see below).

MUIR-WOOD, H. M.; G. F. ELLIOTT AND K. HATAI
1965.   Mesozoic and Cenozoic Terebratellidina *in* Williams *et al.,:* H816-H857, figs. 696-741 (see below).

d'ORBIGNY, A.*
1845.   Molusques *in* M. Ramon de la Sagra. Histoire, Physique, Politique et Naturelle de L'ile de Cuba, pt. 2, vi + 1-376 pp., pls. 1-4, 4bis, 5-28.

1853.   Mollusques *in* M. R. de la Sagra. Same title as above, pt. 2: 1-38.

1847.   Sur les Brachiopodes ou Palliobranches. C. R. Acad. Sci., Paris, *25*(7): 266-269.

PAINE, R. T.
1963.   Ecology of the brachiopod *Glottidia pyramidata* Ecol. Mon., *33*: 255-280 (separate p. 187-213).

PAJAUD, D. AND Y. TAMBAREAU
1970.   Brachiopodes nouveaux du "Sparnacien" des Petities Pyrénées et du Plantaurel. Bull. Soc. Hist. nat. Toulouse, *106*(3-4):312-327.

POURTALÈS, L. F. DE
1867   Contributions to the fauna of the Gulf Stream at great depths. 1st
[1868]. series (1867), Bull. Mus. Comp. Zool. Harvard, *1* (6): 103-120. 2nd series (1868): 121-142.

REEVE, L. A.
1861.   Conchologia Iconica, or figures and descriptions of the shells of molluscous animals. London, 23 vols.

1862.   A revision of the history, synonymy, and geographical distribution of the recent Craniae and Orbiculae. Ann. Mag. Nat. Hist., (3) 7: 126-133.

ROWELL, A. J.
1965.   Inarticulata, *in* Williams, *et al.* :H260-H299, figs. 158-187. See below.

SCACCHI, A. AND R. A. PHILIPPI
1844.   Enumeratio Molluscorum Siciliae, *2*: 18 pls. Halle.

SCHUCHERT, C. AND C. M. LEVENE
1929.   Brachiopoda (Generum et genotyporum index et bibliographia): Fossilium Catalogus, 1, Animalia, Pars 42. Junk, Berlin. 140 pp.

SHIPLEY, A. E.
1883.   On the structure and development of *Argiope*. Mitt. Zool. Stat. Neapel, 4: 494-521.

SOWERBY, G. B.
1847.   The Recent Brachiopoda, Thesaurus Conchyliorum, pts. 6, 7.

*Both volumes cited here with dates 1845 and 1853 have a description of *Discinisca antillarum* by d'Orbigny which is dated 1846, but no reference is given to a volume with the date 1846.

STEINICH, G.
1963.   Drei neue Brachiopoden Gattungen des Subfamilie Cancellothyrinae
         Thomson. Geologie, Berlin, *12* (6): 732-740.

STENZEL, H. B.
1964.   Stratigraphic and paleoecologic significance of a new Danian brachio-
         pod species from Texas. Geol. Rundschau, *54*: 619-631.

STIMPSON, W.
1860.   A trip to Beaufort, North Carolina. Amer. J. Sci., 2nd ser., *39*(87):
         442-445.

THOMSON, J. A.
1915.   A new genus and species of the Thecideinae. Geol. Mag., dec. 6,
         2: 461-464.

1918.   Brachiopoda. Australasian Antarctic Expedition, 1911-1914, Sci. Rept.,
         ser C, Zoology and Botany, *4* (3): 5-76, pls. 15-17.

1926.   A revision of the subfamilies of the Terebratulidae (Brachiopoda).
         Ann. Mag. Nat. Hist., (9) *18:* 523-530.

1927.   Brachiopod morphology and genera (Recent and Tertiary). New Zea-
         land Board Sci. Art, Manual 7, 338 p., 2 pls.

TUNNELL, J. W.
1972.   *Crania* sp. (Brachiopoda) from Texas waters. Texas J. Sci., *23* (4):
         553.

VERRILL, A. E.
1900.   Additions to the Tunicata and Molluscoidea of the Bermudas.
         Trans. Conn. Acad. Arts Sci., *10*: 591, 592, pl. 70.

WAAGEN, W. H.
1882-   Salt Range fossils, part 4 (2) Brachiopoda. Palaeontologia Indica
1885.   Mem., Ser. 13, *1*: 329-270, pls. 25-86.

WEINKAUFF, H. C.
1867.   Die Conchylien des Mittelmeeres, ihre geographische und geognos-
         tische Verbeitung, 2 vols. Cassel.

WILLIAMS, A. *et al.*
1965.   Treatise on Invertebrate Paleontology. R. C. Moore editor. Geol.
         Soc. Amer. and Kansas Univ. Press. 2 parts. i-xxxii, H 1-H927, 746
         figs.

WILLIAMS, A. AND A. J. ROWELL
1965.   Classification *in* Williams, *et al.:* H214-H237 (see above).

ZEZINA, O. N.
1965.   Distribution of the deep water brachiopod *Pelagodiscus atlanticus*
         (King). Okeanologiya (Oceanology, Acad. Sci. USSR) *5*(2):354-358.
         American edition, Amer. Geophys. Union.

1970.   Brachiopod distribution in the Recent ocean with reference to prob-
         lems of zoogeographic zoning. English translation of Pal. Zhurnal,
         2: 1-21. [O raspredelenii brakhiopod v sovremennom okeane v sryazi
         s voprosami zoogeograficheskogo rayonirovaniya].

ZITTEL, K. A. VON
1880.   Handbuch der Palaontologie, R. Oldenbourg, Munchen and Leipzig,
         1(4): 641-722, figs. 473-558.

PLATE 1

Chlidonophora, Argyrotheca and *Pelagodiscus*

Chlidonophora incerta (Davidson): Figure 1.—Ventral, laterally tilted and posterior views, x4, of the interior of the dorsal valve showing the loop, hypotype USNM 550663. For additional views see plate 19. *Locality*: P 1266.

Argyrotheca johnsoni Cooper: Figures 2-7.—2, Ventral view of a young specimen, x4, showing the strong, rounded costae and scalloped anterior, paratype USNM 431003b; 3, anterior view of a young specimen, x4, showing close attachment to the substrate and the strong costation, paratype USNM 431003f; 4, Another young specimen, x4, showing strong costae, paratype USNM 431003c; 5, Young specimen, x4, showing intercalated median rib on the ventral valve, paratype USNM 431003a; 6, 7, Ventral and dorsal views, x4, of the holotype showing strong, rounded costae and intercalated costa in the ventral sulcus, USNM 431003. *Locality*: Johnson-Smithsonian Expedition Station 52.

Pelagodiscus atlanticus (W. King): Figures 8-13.—8, 9, Dorsal view of two specimens attached to coaly material and showing some of the long setae, x4, hypotypes USNM 550539a, b; 10, ventral view, x5, of hypotype USNM 550539b showing a dark patch representing the area of the pedicle opening with coaly material adhering to the pedicle. *Locality*: P338.—11, Dorsal view of a complete specimen in alcohol, x4, hypotype USNM 550729a, showing the long setae. *Locality*: P120.—12, 13, Ventral and dorsal views, x4, of another specimen in alcohol showing the setae, the position of the pedicle (black coaly material adhering) and some of the muscle attachments, hypotype USNM 550730. *Locality*: P1181.

142

PLATE 2

Cryptopora, Crania, Glottidia and *Discradisca*

Cryptopora rectimarginata Cooper: Figures 1-4.—1, 2, Dorsal and ventral views of a specimen in alcohol, x8, showing the lophophore through the thin shell, hypotype USNM 550731; 3, 4, Dorsal and ventral views of another specimen, x8, hypotype USNM 550732. *Locality:* G678.

Crania pourtalesi Dall: Figures 5-11.—5, 6, Exterior, x1, x2, of a dorsal valve showing concentric lines, irregular surface and shape, hypotype USNM 111023: 7, Interior of the preceding dorsal valve, x3, showing muscle scars and small median elevation; 8, Interior of the cemented ventral valve of the preceding specimen, x3, showing muscle and pallial marks. *Locality:* Off Cuba at 260 fathoms (= 476 meters).—9, Exterior of a specimen attached to a pebble, x1, holotype USNM 111022; 10, 11, Exterior and interior of the holotype, x3, showing muscles and septum. *Locality:* Off the Sambos, Florida at 114 fathoms (= 209 meters).

Glottidia pyramidata (Stimpson): Figure 12.—Dorsal view, x2, of a medium size specimen showing long pedicle with attachment at end, hypotype USNM 334748a. *Locality:* Tampa Bay, Florida.

Discradisca aff. *D. antillarum* (D'Orbigny): Figure 13.—Dorsal view, x2, of a dorsal valve having strong radial lines, figured specimen USNM 364180a. *Locality:* Sao Francisco, S. Catharina, Brazil, at 47 fathoms (= 86 meters).

Discradisca antillarum (D'Orbigny): Figures 14-24.—14, 15, Dorsal and side views of a well ornamented specimen, x3, hypotype USNM 64335c; 16-18, Dorsal view, x1, x2 and side view, x2, of another complete specimen, hypotype USNM 64335a; 19-21, Dorsal view x1, and dorsal and ventral views, x3, of the preceding specimen, showing the ventral valve in position with plates (listrium) modifying the foramen; 22-24, Dorsal and ventral views, x3, and side view, x2 of a flatter specimen than the two preceding, hypotype USNM 64335b. *Locality:* Off Jamaica.

144

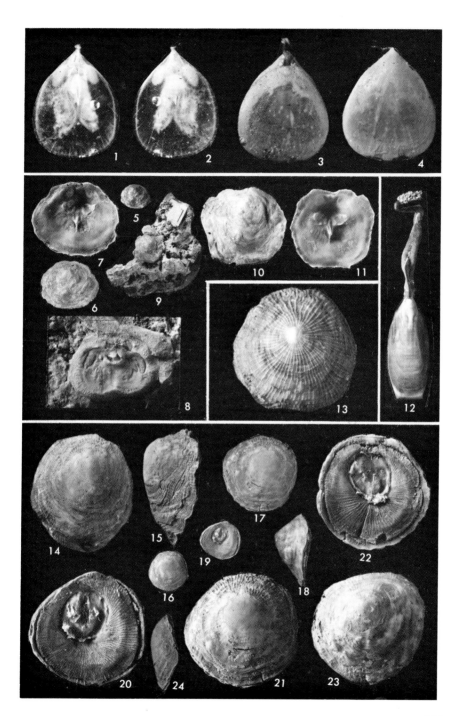

PLATE 3

Argyrotheca, Tichosina, Crania and *Thecidellina*

Argyrotheca bermudana (Dall): Figures 1-5.—Posterior, anterior, side, ventral and dorsal views, x10, hypotype USNM 550531. *Locality*: P855.

Tichosina truncata Cooper, n. sp.: Figures 6-9.—6, Latex impression of the interior of the ventral valve showing muscle and pallial marks, x1, taken from paratype USNM 550587. *Locality*: P594.—7-9, Anterior, side and dorsal views of a complete specimen, x1, holotype USNM 550524a. *Locality*: P584.

Tichosina sp. 7: Figures 10-17.—10-12, Anterior, dorsal and side views of a young specimen, x1, figured specimen USNM 550611b; 13, dorsal interior of the preceding specimen, x1, showing loop; 14-17, Anterior, ventral, side and dorsal views of a larger specimen than the preceding, x1, figured specimen USNM 550611a. *Locality*: OREGON Station 2081.

Crania pourtalesi Dall: Figure 18.—Interior of the dorsal valve, x4, showing muscle callosities, scars and anterior pallial marks, figured specimen USNM 549448a (see plate 4, figs. 18, 19) for attached *Lacazella*. *Locality*: Smithsonian-Johnson Expedition Station 52.

Thecidellina cf. *T. barretti* (Davidson): Figures 19-21.—19. Interior of the ventral valve, x20. showing hemispondylium, hypotype USNM 550540a; 20, Dorsal valve, counterpart to preceding ventral valve, x20, showing squarish cardinal process, and slender median ridge, strongly contrasting with the structure of *Lacazella* (see plate 3); 21, Exterior of a complete specimen, x20, hypotype USNM 550540b. *Locality:* P630.

Thecidellina barretti (Davidson): Figures 22-26.—22, 23, Ventral valve tilted and in dorsal view, x6, showing interarea with its flat cover, hemispondylium and thick teeth, hypotype USNM 550541; 24, 25, interior of dorsal valve belonging to the preceding and showing the slender median septum, x6, in ventral and anteriorly tilted views; 26, the complete specimen, x6, with the dorsal valve in place. *Locality*: NGS-MRF-33.

Tichosina? *bartletti* (Dall): Figures 27, 28.—Interior of the dorsal valve showing the long loop in ventral and side views, x2, with its elevated ridges (crural bases) along the inside margin of the hinge plates, hypotype USNM 549393a. *Locality*: Johnson-Smithsonian Expedition Station 102.

146

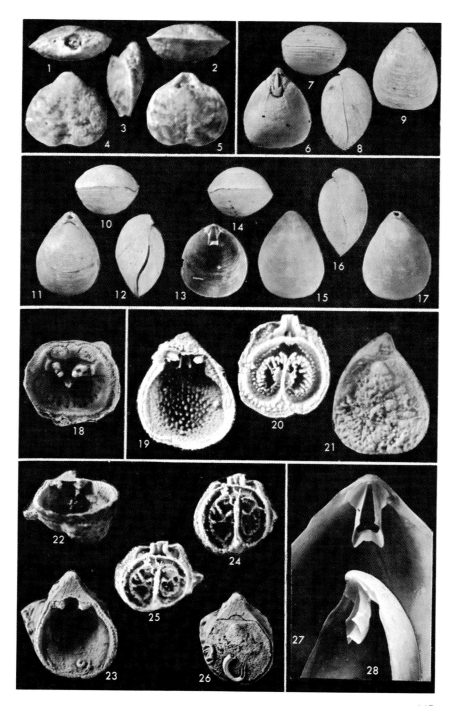

PLATE 4

Dallina and *Lacazella*

Dallina floridana (Pourtalès): Figures 1-4.—1, 2, Ventral and dorsal views of an immature specimen in alcohol showing lophophore, x8, hypotype USNM 550616; 3, 4, Interior of the ventral and dorsal valves of the preceding specimen, x8, showing dental plates, crura and initial pillar of the median septum. *Locality*: P988.

Lacazella mediterranea (Risso): Figures 5-11.—5-7, Dorsal, anterior and side views, x4, of a complete specimen, hypotype USNM 274154a. *Locality*: Mediterranean.—8, 9, Interior of the two ventral valves tilted to show the hemispondylium and strongly granulose surface, x5, hypotypes USNM 173494b, c; 10, 11, Interior of the two dorsal valves, x6, showing spinose ascending and descending apparatus, hypotypes USNM 173594c and d (fig. 11 is counterpart to fig. 9). *Locality*: Port Vendres, France.

Lacazella caribbeanensis Cooper, n. sp.: Figures 12-19.—12, Dorsal view of a specimen attached to *Argyrotheca,* x5, showing dorsal valve in place, holotype USNM 549449a; 13, Interior of the ventral valve of the holotype showing hemispondylium, x12; 14, 15, Interior of the dorsal valve of the holotype with lophophore filaments and with filaments removed to show inner smooth skeleton, ca x15; 16, 17, Ventral interior tilted and in dorsal view showing hemispondylium of another specimen, x10, paratype USNM 549449b; 18, 19, Anterior and dorsal views of two specimens attached to the dorsal valve of *Crania* (see plate 3, fig. 18), x4, paratype USNM 549448a. *Locality*: Smithsonian-Johnson Expedition Station 52.

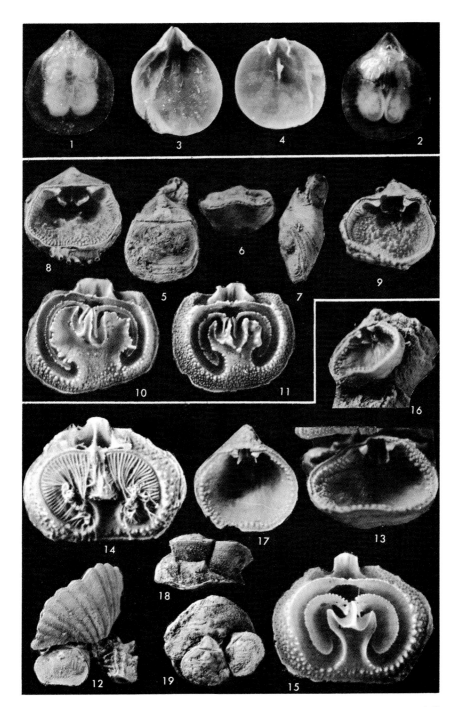

PLATE 5

Cryptopora, Tichosina and Erymnia

Cryptopora rectimarginata Cooper: Figures 1-8.—1-3, Side, dorsal and anterior views, x6, showing rectimarginate commissure and wide foramen, holotype USNM 274143a; 4, 5, Interior of the ventral valve, specimen tilted to the rear and in dorsal view, x6, to show deltidial plates and strong dental plates. paratype USNM 274143c; 6, 7, laterally tilted and ventral views, x8, of the preceding specimen showing the prominent, high median septum and maniculifer crura; 8, Dorsal view of the posterior of a complete specimen, x15, showing elaborate alate deltidial plates and apical plate in apex, paratype USNM 274143d. *Locality*: EOLIS Station 340.

Tichosina plicata Cooper, n. sp.: Figures 9-15.—9-11, Side, anterior and dorsal views, x1, of a specimen with strongly uniplicate anterior commissure, holotype USNM 550607a; 12, posterior of the pedicle valve, x2, of a paratype USNM 550607e; 13, 14, Partial side and ventral views of the dorsal valve interior, x2, showing short, wide loop, paratype USNM 550607e (counterpart to the preceding); 15, Latex impression, of the ventral valve interior of the preceding, x1, showing muscle scars. *Locality*: OREGON Station 5624.

Erymnia angusta Cooper, n. sp.: Figures 16-24.—16-20, Dorsal, anterior, ventral, posterior and side views, x1, of the holotype, USNM 559608a; 21, 22, Side and ventral views of the loop, x2, of the holotype showing long, narrow loop and loop supports; 23, Posterior of the ventral valve of the holotype, x2, showing the small, strongly labiate foramen; 24, Latex impression of the ventral interior of the holotype, x1, showing muscle scars. *Locality*: SILVER BAY Station 3494.

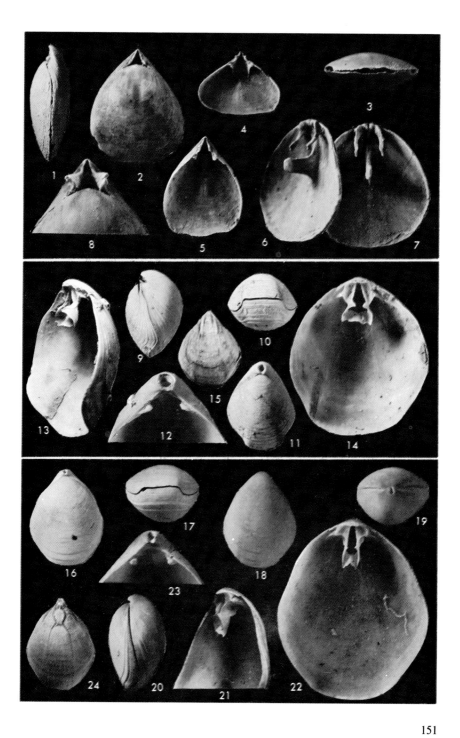

151

PLATE 6

Tichosina

Tichosina bullisi Cooper, n. sp.: Figures 1-8.—1-5, Anterior, ventral, side, posterior and dorsal views, x1, of the holotype, USNM 550609a; 6, Posterior of the ventral valve of the holotype showing teeth and foramen, x2; 7, Rubber impression of the interior of the ventral valve of the holotype, x1, showing muscle scars; 8, posterior of the dorsal valve of the holotype, x2, showing the broad ribboned, stout loop. *Locality*: OREGON Station 3608.

Tichosina expansa Cooper, n. sp.: Figures 9-16.—9-13, Dorsal, posterior, ventral, side and anterior views of the holotype, x1, USNM 550610; 14, Rubber impression, x1, of the ventral interior of the holotype showing muscle scars; 15, Posterior of the ventral valve of the holotype showing the foramen, x2; 16, Posterior of the dorsal valve of the holotype, x2, showing the narrow loop with its broad and narrowly folded transverse ribbon. *Locality*: SILVER BAY Station 3499.

153

PLATE 7

Tichosina

Tichosina subtriangulata Cooper, n. sp.: Figures 1-17.—1-4, Ventral dorsal, anterior and side views, x1, of the holotype, USNM 226290-41; 5-8, Anterior, ventral, dorsal and side views of a large paratype, x1, USNM 226290-10; 9-11, Anterior, side and dorsal views of a small individual, x1; paratype USNM 226290-23; 12-15, Ventral, anterior, side and dorsal views, x1, paratype USNM 226290-5; 16, Interior of the dorsal valve showing narrow loop with narrow, deeply concave hinge plates bordered by the crural bases, x2, paratype USNM 226290a; 17, Latex impression of the ventral valve interior, x1, showing muscle scars of paratype USNM 226290e. *Locality*: FISHHAWK Station 6070.

Tichosina rotundovata Cooper, n. sp. Figures 18-32.—18-22, Ventral anterior, side, posterior and dorsal views, x1, of a paratype USNM 550526a; 23-27, side, anterior, posterior, ventral and dorsal views, x1, of the holotype USNM 550526f; 28, 29, Impressions of the interior of the ventral and dorsal valves, x2, showing muscle scars and pallial trunks, paratype USNM 550526g, h; 30, 31, Interior of the dorsal valve, x1, x2, showing long narrow loop, paratype USNM 550526a; 32, Posterior of the ventral valve, x2, of the preceding specimen showing symphytium and teeth. *Locality*: G482.

154

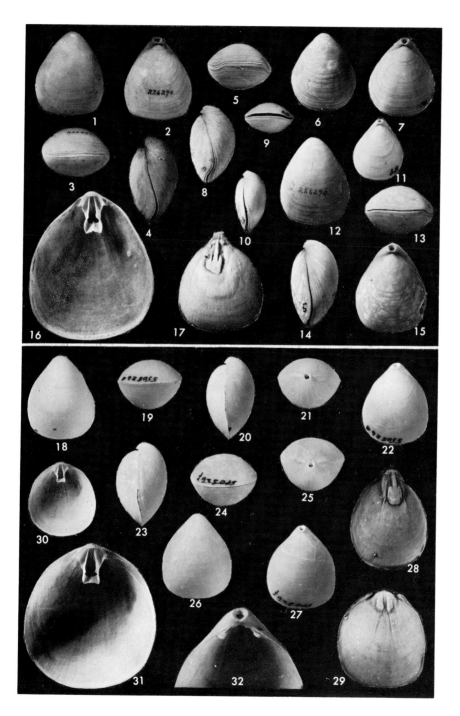

PLATE 8

Tichosina

Tichosina martinicensis (Dall): Figures 1-9.—1-5, Side, dorsal, posterior, ventral and anterior views, x1, of the holotype, USNM 64255; 6, 7, Interior of the dorsal valve of the holotype x1, x2, showing loop; 8, 9, Interior of the ventral valve of the holotype, x1, x2 showing teeth, pallial marks and body wall. *Locality*: BLAKE Station 193.

Tichosina bahamiensis Cooper, n. sp.: Figures 10-26.—10-13, Dorsal view, x1, and dorsal, side and anterior views, x2, holotype USNM 87378a; 14-17, dorsal view, x1, and side, anterior and dorsal views, x2, of an unusually large specimen, paratype USNM 87378b; 18-22, Dorsal view, x1, and dorsal, anterior, ventral and side views, x2, of a slender specimen, paratype USNM 837378c; 23, 24, impressions of the interior of the dorsal and ventral valve, x2, paratype 87378f; 25, Interior of the dorsal valve, x3, showing long narrow loop, paratype USNM 837378g; 26, Posterior of the ventral valve of the preceding specimen, x3. *Locality*: ALBATROSS Station 2655.

156

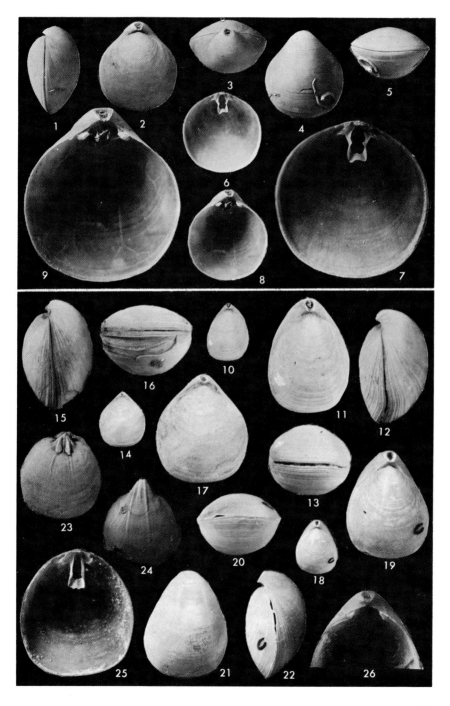

PLATE 9

Tichosina

Tichosina obesa Cooper, n. sp.: Figures 1-17.—1-3, Dorsal, anterior and side views, x1, of a characteristic specimen, holotype USNM 550595a; 4-6, Ventral, side and anterior views, x1, of a fairly wide specimen, paratype USNM 550585d; 7-9, Anterior, dorsal and side views of a narrow specimen, x1, paratype USNM 550585b; 10, 11, Interior of the ventral and dorsal valve, showing wide loop, x2, paratype USNM 550585e. *Locality*: P734.—12-14, Anterior, dorsal, and side views of a large specimen, x1, paratype USNM 550514b; 15-17, Anterior, side and dorsal views, x1, of a large specimen, paratype USNM 550514a. *Locality*: P737.

Tichosina erecta Cooper, n. sp.: Figures 18-27.—18-21, Side, anterior, ventral and dorsal views, x1, of the holotype showing short pedicle, USNM 550525a; 22-24, Anterior, side and dorsal views of a specimen smaller than the preceding, x1, paratype USNM 550525b; 25, Interior of the dorsal valve, x2, showing long slender loop with elevated margins along the hinge plates, paratype USNM 550525c; 26, 27 Interior of the dorsal valve x3, x2, showing long slender loop, paratype USNM 550525d. *Locality*: G694.

Tichosina solida Cooper, n. sp.: Figures 28-34.—28-30, Side, anterior and dorsal views, x1, holotype USNM 549433a; 31, 32, Interior, x1, and posterior x2, of the ventral valve showing foramen, paratype USNM 549433b; 33, 34, Interior of the dorsal valve belonging to the preceding specimen, x1, x2, showing narrow loop. *Locality*: Eolis Station 1, Sand Key, Florida.

158

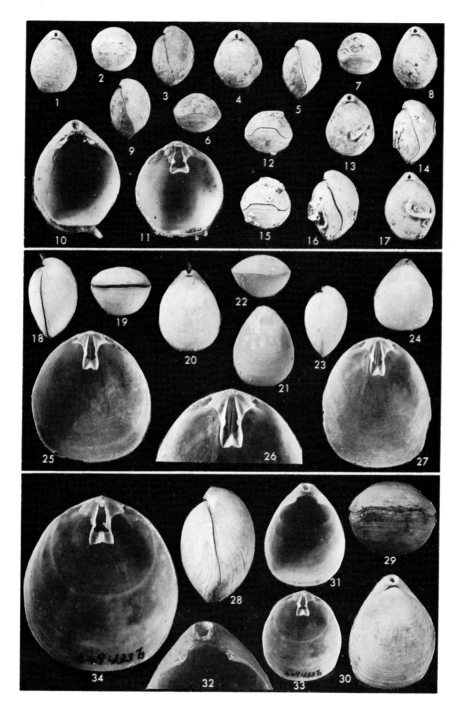

159

PLATE 10

Tichosina

Tichosina abrupta Cooper, n. sp.: Figures 1-10.—1-5, Dorsal view, x1, and ventral, anterior, side and dorsal views, x2, of the holotype, USNM 550599; 6, Latex impression of the interior of the ventral valve of the holotype, x2; 7, Posterior of the holotype, x2, showing symphytium and teeth; 8, Interior of the dorsal valve, x2, showing, short, stout loop. *Locality*.: Eolis Station 1.—9,10, Interior of the dorsal and ventral valves of a thick-shelled paratype, x2, USNM 336848. *Locality*: EOLIS Station 31.

Tichosina? bartletti (Dall): Figures 11-17.—11-13, Anterior, dorsal and side views, x1, of the holotype, USNM 110852; 14, Latex impression of the interior of the ventral valve of the holotype, x1, showing muscle scars and pallial marks; 15, Posterior of the ventral valve of the holotype, x2, showing symphytium; 16, 17, Ventral,and side views of the dorsal valve interior of the holotype, x2, showing loop with broad crural bases forming an elevated margin to the hinge plates. *Locality*: Barbados.

160

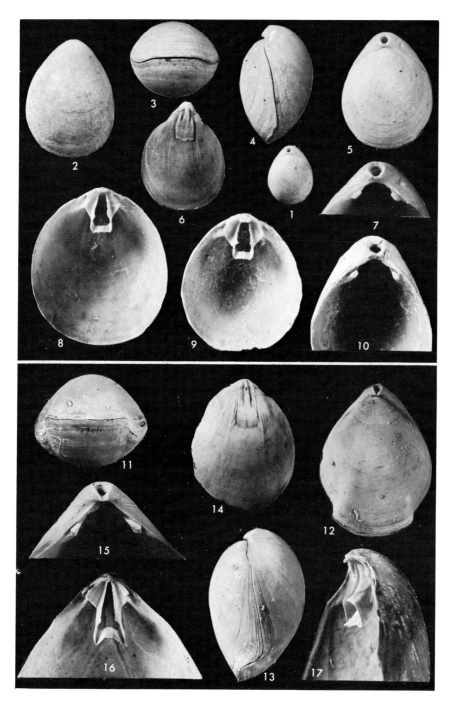

PLATE 11

Tichosina, Dallina and *Abyssothyris?*

Tichosina floridensis Cooper, n. sp.: Figures 1-15.—1-3, Anterior, side and dorsal views of specimen of medium size, x1, paratype USNM 64248a; 4-6, Side, dorsal and anterior views of another paratype, x1, USNM 64248b; 7, Interior of the ventral valve, x2, showing teeth and strong pallial marks, USNM 64248d; 8, 9, Interior of the dorsal valve of the preceding paratype, x1, x2, showing the loop. *Locality*: Off Havana, Cuba.—10-14, Ventral, side, anterior, posterior and dorsal views, x1, of a large specimen, holotype USNM 550738a; 15, Interior of a paratype showing the loop, x2, USNM 550738b. *Locality*: OREGON Station 955.

Tichosina bartschi (Cooper): Figure 16.—Dorsal view of the holotype, x1, showing narrowly rounded anterior and maximum width at midvalve, holotype USNM 431002.

Abyssothyris? sp. 1: Figures 17-19.—Anterior, side and dorsal views of a small specimen, x1, hypotype USNM 110853. *Locality*: Albatross Station 2035.

Dallina floridana (Pourtalès): Figures 20-27.—20-24, Anterior, posterior, side, ventral and dorsal views, x1, of a large typical specimen hypotype USNM 550583. *Locality*: G246.—25-27, Ventral, posterior and laterally tilted views, x1, of the large, thick loop, hypotype USNM 550582. *Locality*: SILVER BAY Station 2416.

162

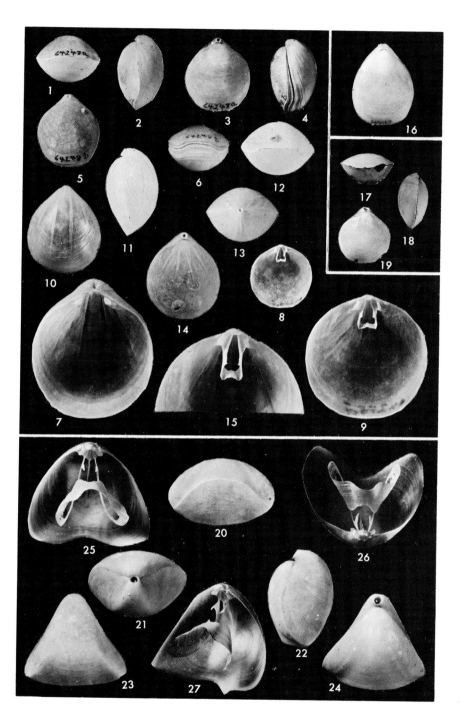

PLATE 12

Tichosina and Erymnia

Tichosina truncata Cooper, n. sp.: Figures 1, 2.—Interior of the dorsal and ventral valves, x2, showing narrow, broad-ribboned loop, holotype USNM 550524a. (See plate 3, figures 6-9 for exterior views of the holotype.) *Locality*: P584.

Erymnia muralifera Cooper, n. sp.: Figures 3, 4.—Anterior and side views of an imperfect dorsal valve, x3, showing the strengthening partitions on the under (dorsal) side of the loop, paratype USNM 550624. *Locality*: G704.

Tichosina sp. 8. Figures 5-7.—Side, anterior and dorsal views of a complete specimen, x1, figured specimen USNM 550546a. Formerly identified as *Tichosina cubensis* (Pourtales). *Locality*: BLAKE Station 167.

Tichosina sp. 1. Figures 8-12.—Dorsal, posterior, side, anterior and ventral views of the only specimen, x1, figured specimen USNM 550665. Note widely oval outline and narrowly curved anterior commissure. *Locality*: OREGON Station 6715.

Tichosina elongata Cooper, n. sp.: Figures 13-17.—Ventral, anterior, side, posterior and dorsal views of this long, slender species, x1, holotype USNM 550664. *Locality*: COMBAT Station 450.

Tichosina dubia Cooper, n. sp.: Figure 18.—Interior of the ventral and dorsal valves, x2, showing the stout, expanding loop, paratype USNM 550614a. *Locality*: OREGON II Station 10513.

Tichosina floridensis Cooper, n. sp.: Figures 19-24.—19-23, Ventral, anterior, side, dorsal and posterior views, x1, of a large individual, paratype USNM 550737-15; 24, Interior of the dorsal valve, x2, showing stout loop, paratype USNM 550737e. *Locality*: OREGON Station 1025.

164

PLATE 13

Erymnia

Erymnia muralifera Cooper, n. sp.: Figures 1-22.—1-5, Dorsal posterior, side, anterior and ventral views, x1, of a specimen of medium size paratype USNM 550523; 6, 7, Interior of the ventral and dorsal valves of the preceding specimen, x1; 8, 9, Posterior of the preceding dorsal valve, x3, tilted slightly laterally and in ventral views showing the hinge plate supports. *Locality*: G706.—10-14, Ventral side, dorsal, posterior and anterior views of a complete specimen, x1, paratype USNM 550521a. *Locality*: G691.—15-17, Side anterior and dorsal views of a large specimen with numerous epifauna, x1, paratype USNM 550578; 18, Latex impression of the interior of the ventral valve of the preceding specimen, x1; 19, Interior of the dorsal valve of the preceding specimen, x1; 20, posterior of the preceding specimen, x3, showing supports of the hinge plates. *Locality*: P991.—21, 22, Lateral and posteriorly tilted views, x3, showing narrow body cavity, paratype USNM 550522. *Locality*: G646.

167

PLATE 14

Erymnia and *Tichosina*

Erymnia muralifera Cooper, n. sp.: Figures 1-10.—1-5, Dorsal, side, ventral, posterior and anterior views, x1, of the holotype USNM 550520; 6, Interior of the ventral valve, x1, of the holotype; 7, Interior of the dorsal valve of the holotype, x1, showing the loop; 8-10, Ventral, laterally tilted and posteriorly tilted views of the dorsal valve of the holotype, x3, showing loop and supporting plates. *Locality*: G695.

Tichosina labiata Cooper, n. sp.: Figures 11-20.—11-15, Dorsal, side, anterior, ventral and posterior views, x1, holotype USNM 550577a; 16, 17, Interior of the ventral valve of the preceding specimen; x1, x2, showing teeth and symphytium; 18, 19, Interior of the dorsal valve of the preceding specimen, x1, x2, showing fairly wide loop with elevated borders on the hinge plates; 20, Impression of the interior of the ventral valve of the preceding specimen, x1½, showing muscle scars and pallial trunks. *Locality*: P876.

168

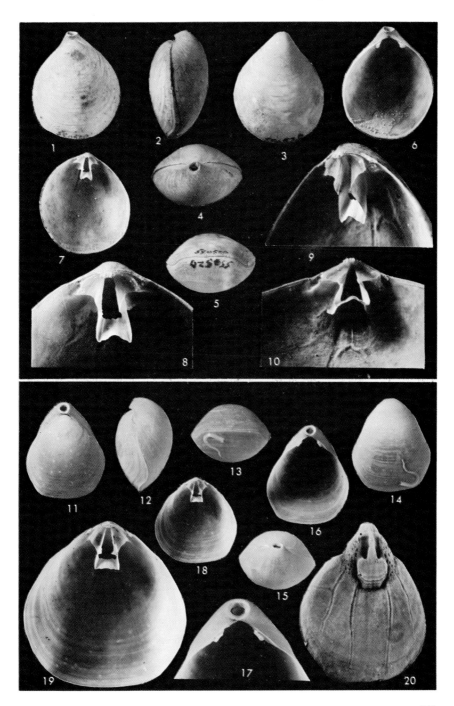

169

PLATE 15

Dyscolia

Dyscolia wyvillei (Davidson): Figures 1-10.—1, Dorsal view of a complete specimen, x1, showing triangular shape, hypotype USNM 549271; 2, Interior of the dorsal valve of the preceding specimen, x1, showing the short anteriorly rounded loop; 3, Loop of the preceding dorsal valve, x2 illustrating its simple form. *Locality*: Northwest coast of Africa.—4, 5, Exterior and interior views, x1, of the dorsal valve showing hinge plates, hypotype USNM 550621c; 6, Exterior of another dorsal valve showing trace of radial lines, x1, hypotype USNM 550621e; 7, The preceding specimen enlarged, x2, to show the radial lines on the exterior; 8-10, Exterior, interior and posterior views, x1, of the ventral valve showing the teeth, thick symphytium and large labiate foramen, hypotype USNM 550621b. *Locality*: P1262.

170

PLATE 16

Eucalathis, Argyrotheca and *Notozyga*

Eucalathis cubensis Cooper, n. sp.: Figures 1-8.—1-5, Dorsal view, x1, and anterior, dorsal, side and ventral views, x3, holotype USNM 110848; 6, Interior of the ventral valve, x3, of the holotype; 7, 8, Ventral and laterally tilted views of the dorsal valve showing the characteristic pointed loop, x6, of the holotype. *Locality*: BLAKE Station 16.

Eucalathis floridensis Cooper, n. sp.: Figures 9-11.—Ventral, side and dorsal views, x3, of the holotype, USNM 107526. *Locality*: Pourtales platform, Florida Keys.

Argyrotheca schrammi (Crosse and Fischer). Figures 12-16.—Ventral, anterior, posterior, dorsal and side views of a complete specimen, x10, hypotype USNM 550601. *Locality*: NGS-MRF-33.

Notozyga lowenstami Cooper, n. sp.: Figures 17-31.—17, Dorsal view of a small specimen, x10, paratype USNM 550591e; 18-22, Dorsal, side, anterior, posterior, and ventral views, x10, holotype USNM 550591a; 23, Interior of a large ventral valve, x10, showing interareas, large teeth and rudimentary deltidial plates, paratype USNM 550591d; 24, 25, Interior and exterior, x10, x8, showing lophophore and exterior ornament, paratype USNM 550591c; 26, Exterior of a ventral valve, x10, showing beaded ornament, paratype USNM 550591b; 27-29, Anterior, ventral and laterally tilted views of the dorsal valve of the holotype, x10, showing loop with its receding transverse ribbon; 30, Ventral view of the preceding specimen x15, showing loop in greater detail; 31, Interior of the dorsal valve of paratype USNM 550591b, x17.5 showing the lophophore. *Locality*: California Institute of Technology Station 1494.

172

PLATE 17

Terebratulina and *Megerlia*

Terebratulina latifrons Dall: Figures 1-13.—1-6, Dorsal view, x1, and ventral, posterior, anterior, side and dorsal views, x2, of a characteristic specimen, hypotype USNM 550579a; 7-12, Dorsal view, x1, and anterior, dorsal, posterior, side and ventral views, x2, of a subcarinate specimen, hypotype USNM 550579b. *Locality*: G983.—13, Interior of the dorsal valve showing the stout loop with complete ring, x3, hypotype USNM 550519. *Locality*: G947.

Megerlia echinata (Fischer and Oehlert): Figures 14-22.—14-16, Side anterior and dorsal views of a complete specimen, x2, hypotype USNM 550527; 17, Ventral view of the same specimen, x3, showing the small spines; 18, Posterior of the ventral valve, x6, of the preceding specimen showing the teeth and median ridge; 19-22, Side, anterior, ventral and posterior views of the dorsal valve interior of the same hypotype showing the loop, x6. *Locality*: P739.

174

PLATE 18

Tichosina, Platidia and *Abyssothyris?*

Abyssothyris? parva Cooper, n. sp.: Figures 1-7.—1-5, Dorsal view, x1, and side, anterior, ventral and dorsal views, x3, holotype USNM 550593a; 6, Posterior of the ventral valve showing foramen and symphytium, x4, paratype USNM 550593b; 7, Interior of the dorsal valve showing the anteriorly rounded loop, x4, belonging to the preceding ventral valve. *Locality*: ATLANTIS Station 266-2.

Tichosina sp. 3: Figures 8-11.—Anterior, side, ventral and dorsal views of a complete specimen, x1, figured specimen USNM 64260. *Locality*: BLAKE Station 253.

Platidia davidsoni (Deslongchamps): Figures 12-22.—12-14, Dorsal, side and ventral views of a complete specimen, x5, hypotype USNM 550545a; 15, Exterior of the ventral valve of the hypotype, x10, showing the pustulose ornament. *Locality*: G304.—16, Interior of the ventral valve showing teeth and short median ridge, x5, hypotype USNM 550544a; 17, Interior of the dorsal valve of the preceding specimen showing the loop, x5; 18-22, Posterior, anterior, partial anterior, side and ventral views of the same dorsal valve, x10, showing the loop in greater detail. *Locality*: G190.

176

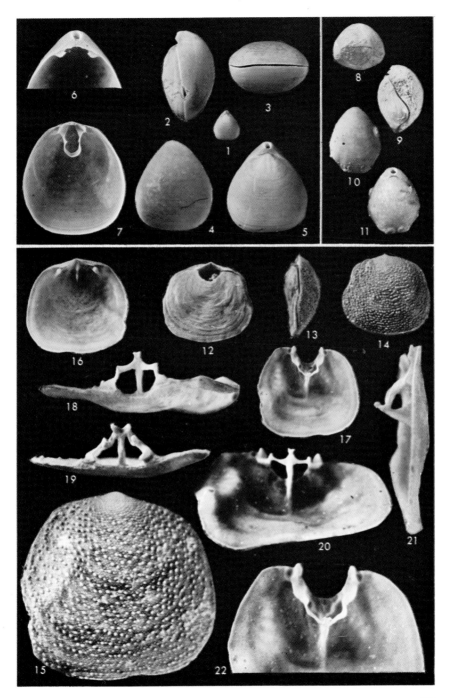

177

PLATE 19

Glottidia, Tichosina and Chlidonophora

Glottidea audebarti (Broderip): Figure 1.—Ventral view of a large specimen, x1, showing the long pedicle, hypotype USNM 550536. The specimen is a bright grass green in the anteromedian part. *Locality*: P572.

Tichosina dubia Cooper, n. sp.: Figures 2-7.—2-5, Dorsal, anterior, ventral and side views, x1, of a complete specimen, holotype USNM 64261; 6, 7, Interior of the dorsal valve of the preceding specimen, x2, showing wide loop. *Locality*: BLAKE Station 147.

Chlidonophora incerta (Davidson): Figures 8-22.—8-12, Dorsal, side, anterior, posterior and ventral views, x3, hypotype USNM 550586a; 13, Dorsal view of the preceding hypotype, x1; 14-19, Dorsal view, x1, and anterior, posterior, side, dorsal and ventral views of another hypotype, x3, USNM 550586b. *Locality*: G325.—Figure 20. Interior of the dorsal valve showing the subplectolophous lophophore, x8, hypotype USNM 550602b. *Locality*: OREGON 2574.—21, 22. Interior of the dorsal and ventral valves, x4, showing loop in the former and interarea and teeth in the latter, hypotype USNM 550584a. *Locality*: P1138.

178

PLATE 20

Abyssothyris and *Platidia*

Abyssothyris atlantica Cooper, n. sp.: Figures 1-10.—1-3, Dorsal, side and anterior views, x1, of the holotype, USNM 550592a; 4-7, Anterior, dorsal, ventral and side views of the holotype, x3, showing sulcate anterior commissure, small pedicle and strong growth lines; 8, Side view of the interior, x4, showing the tightly rolled lophophore, paratype USNM 550592c; 9, Interior of the dorsal valve, x4, showing anteriorly rounded loop, paratype USNM 550592b; 10, Posterior of the ventral valve belonging to preceding dorsal valve showing foramen and symphytium, x4. *Locality*: HYDRO Station 5809.

Platidia anomioides (Scacchi and Philippi): Figures 11-19.—11-13, Ventral, dorsal and side views, x5, showing smooth surface and foramen shared by both valves, hypotype USNM 550528a; 14, Posterior of the ventral valve interior showing teeth and posterior median ridge, x10, hypotype USNM 550528b; 15, Interior of a dorsal valve showing bilobed lophophore, x10, hypotype USNM 550528c; 16-19, Anterior, side, laterally tilted and ventral views of the dorsal valve, x10, showing the loop, hypotype USNM 550528b (dorsal counterpart to ventral valve, figure 14). *Locality*: P861.

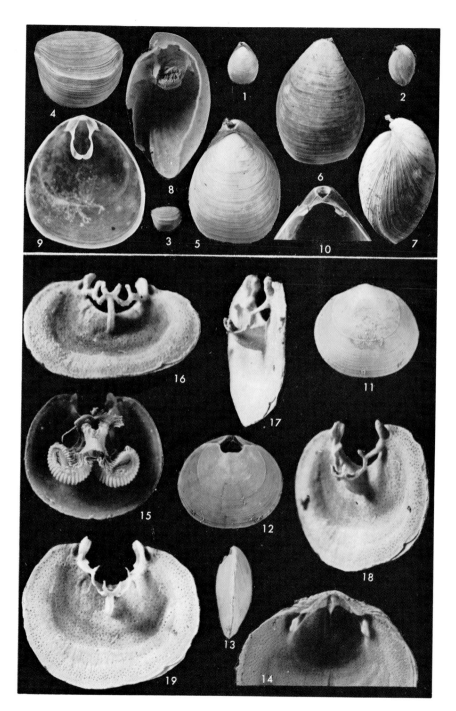

PLATE 21

Tichosina and *Argyrotheca*

Tichosina dubia Cooper, n. sp.: Figures 1-5.—1-3, Dorsal, anterior and side views, x1, showing uniplicate anterior commissure, paratype USNM 550603; 4, 5, Interior of the ventral and dorsal valves of the preceding specimen, x2, showing large foramen, and stout loop. *Locality*: P658.

Tichosina plicata Cooper, n. sp.: Figures 6-11.—6-9, Side, anterior, dorsal and ventral views, x1, showing uniplicate anterior commissure, figured specimen 550588; 10, 11, Interior of the ventral and dorsal valves, x2, of the same specimen. *Locality*: P838.

Argyrotheca rubrocostata Cooper, n. sp.: Figures 12-27.—12-16, Ventral, anterior, dorsal, side and posterior views, x4, paratype USNM 550530a; 17, Interior of the dorsal valve showing the lophophore, x5, paratype USNM 550530b. *Locality:* P629. 18-22, Anterior, posterior, dorsal, side and ventral views of a large adult, x4, holotype USNM 550529a; 23, Dorsal view of the holotype, x1;24-26, Posterior, ventral and side views, x4, of the dorsal valve interior showing the discontinuous loop and serrate median septum, paratype USNM 550529b; 27, Interior of the ventral valve belonging to the preceding dorsal valve, x4, showing wide teeth and strong median septum. *Locality:* P630.

182

183

PLATE 22

Tichosina and *Argyrotheca*

Tichosina cubensis (Pourtalès): Figures 1-8.—1-3, Anterior, dorsal and side views, x1, of an adult specimen, hypotype USNM 109748a. *Locality*: Gulf of Florida.—4-6, Anterior, side, and dorsal views of a large adult, x1, showing sinuous lateral commissure and nearly rectimarginate anterior commissure, hypotype USNM 149405. *Locality*: Straits of Florida.—7.8, Posterior of the dorsal valve interior, dorsal and side views, x2 showing the loop, with its elevated border to the hinge plates, hypotype USNM 110856. *Locality:* FISHHAWK Station 7283.

Argyrotheca barrettiana (Davidson): Figures 9-21.—9-13, Posterior, side, anterior, ventral and dorsal views of a young specimen, x4, hypotype USNM 550604a; 14-18, Dorsal, posterior, anterior, side and ventral views, x4, of a large adult, hypotype USNM 550604b; 19, The preceding specimen opened to show the interior, x4, the large and thick dorsal septum and the loop partially cemented to the valve floor. *Locality*: P389.—20, 21, Interior of the ventral and dorsal valves of a young specimen, x4, showing septum in ventral valve and discontinuous loop in dorsal valve, hypotype USNM 550605. *Locality*: G983.

184

PLATE 23

Argyrotheca and *Megathiris*

Argyrotheca sp. 3: Figs. 1-5.—Ventral, anterior, side, posterior and dorsal views of a complete individual, x4, figured specimen USNM 550740. *Locality*: GERDA-VII-1958.

Argyrotheca barrettiana (Davidson): Figs. 6, 7.—Posterior and ventral views of a large individual, x4, showing pedicle fibers penetrating a fragment of coral, hypotype USNM 550747a. *Locality*: P1434.

Megathiris detruncata (Gmelin): Figs. 8-19.—8-12, Ventral, posterior, anterior, dorsal and side views, x4, of a complete individual, hypotype USNM 199368a; 13-17, Dorsal, anterior, posterior, ventral and side views of another complete specimen, x4, hypotype USNM 199368b; 18, 19, Interior of the ventral and dorsal valves, x6, of the preceding specimen. *Locality*: Guadeloupe?

Argyrotheca sp. 2: Figs. 20-24.—Posterior, dorsal, side, anterior and ventral views, x4, of a large specimen, figured specimen USNM 550742. *Locality*: P630.

Argyrotheca sp. 1: Figs. 25-29—Posterior, anterior, dorsal, side and ventral views, x4, of a complete specimen, figured specimen USNM 550741. *Locality*: P1434.

186

PLATE 24

Tichosina and *Argyrotheca*

Tichosina sp. 4: Fig. 1.—Dorsal view, x1, of a large species, externally reminiscent of the extinct *Terebratula,* figured specimen USNM 550667. *Locality*: OREGON 4938.

Argyrotheca rubrotincta (Dall): Figs. 2-11.—2-6, Posterior, side, dorsal, anterior and ventral views, x4, of the lectotype USNM 110962a; 7, 8, Dorsal and ventral interiors of the lectotype, x6; 9, Dorsal view of an immature specimen showing the strongly serrate anterior margin, x4, paratype USNM 110962b. *Locality*: Tortugas, Gulf of Mexico.—10, 11, Anterior and dorsal views, x4, of a young specimen attached to a coralline fragment, x4, hypotype USNM 82923a. *Locality:* Florida Keys.

Argyrotheca lutea (Dall): Figs. 12-28.—12-16, Dorsal, posterior, ventral, anterior and side views, x4, of a complete specimen, hypotype USNM 32924a; 17, 18, Exterior and interior of a large ventral valve, x4, showing deep pits in median septum, hypotype USNM 32924b; 19, 20, Exterior and interior of the dorsal valve, x4, of the preceding specimen, showing greatly thickened median region. *Locality:* 43 fathoms. Tortugas, Florida.—21-25, Ventral, dorsal, anterior, posterior and side views, x4, of a complete paratype USNM 110963b; 26, 27, Side and ventral views of the dorsal valve interior, x4, showing serrated median septum and loop, lectotype USNM 110963a; 28, Interior of the ventral valve, x4, showing pits in median septum, paratype USNM 110963c. *Locality:* 30 fathoms, Tortugas, Gulf of Mexico.

188

189

PLATE 25

Terebratulina, Argyrotheca and *Tichosina*

Terebratulina cailleti Crosse: Figures 1-16.—1, 2, Side view of a specimen attached to a small pebble, x1, x2, hypotype USNM 550542a; 3-8, Dorsal view, x1, and dorsal, anterior, side, posterior and ventral views, x2, of an average specimen, hypotype USNM 550542b. *Locality*: P929.—9-14, Dorsal view, x1, and posterior, side, anterior, ventral and dorsal views, x2, of an unusually large specimen, hypotype USNM 550543. *Locality*: P600.—15, Interior of the ventral valve, x2, hypotype USNM 550576; 16, Interior of the dorsal valve of the preceding specimen, x2, showing thick loop with crural processes forming a ring. *Locality*: P739.

Argyrotheca crassa Cooper, n. sp.: Figures 17-22.—17, Dorsal view, x1, of a complete specimen, holotype USNM 550581; 18-22, Posterior, dorsal, anterior, side and ventral views of the holotype, x4. *Locality:* P857.

Tichosina solida Cooper, n. sp.: Figures 23-30.—23-27, Dorsal, ventral, posterior, side and anterior views of a complete specimen, x1, paratype USNM 550619; 28-30, Interior of the dorsal valve of the preceding paratype, showing the aberrant loop in partial left and right views (reader's) and in ventral view to show abnormal development, all x2. *Locality*: G1029.

190

PLATE 26

Macandrevia and *Tichosina*

Macandrevia novangliae Dall: Figures 1-11.—1-5, Posterior, anterior, side, ventral and dorsal views, x1, of the holotype USNM 78069; 6, Dorsal exterior, enlarged, x1½, better to show the concentric lines and open delthyrium. *Locality*: ALBATROSS Station 2682.—7, Ventral view of another specimen attached to a pebble, x1, hypotype USNM 49068a; 8, posterior of the ventral valve, x3, showing lack of deltidial plates, hypotype USNM 49068b; 9, Interior of the dorsal valve showing cardinalia, x3, counterpart to the preceding specimen; 10, Interior view of the preceding specimens, x2, tilted to show the hinge plates meeting the valve floor; 11, Dorsal valve showing free loop, x2, hypotype USNM 550625. *Locality*: ALBATROSS Station 2208.

Tichosina pillsburyae Cooper, n. sp.: Figures 12-27.—12-16, Anterior, ventral, side dorsal and posterior views of a young specimen showing small foramen, x1, paratype USNM 550622c; 17, Posterior of the dorsal valve interior of the preceding specimen, x2, showing the loop; 18, Latex impression of the interior of a ventral valve, x1, showing the muscle scars and the main pallial trunks, paratype USNM 550622a; 19-23, Anterior, dorsal, side, posterior and ventral views of a large specimen, x1, holotype USNM 550622b, showing strongly uniplicate anterior commissure; 24, 25, Interior of the ventral valve of the holotype x1, x2, showing thickened interior and foramen; 26, 27, Interior of the dorsal valve of the holotype, x1, x2, showing characteristic *Tichosina* loop. *Locality*: P1387.

192

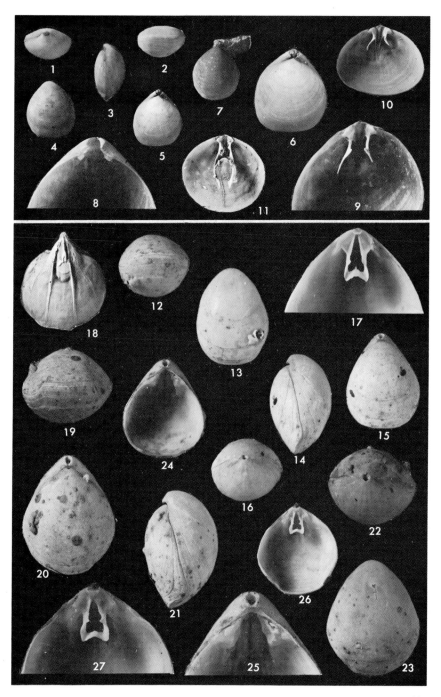

PLATE 27

Glottidia, Platidia, Cryptopora and *Tichosina*

Glottidia pyramidata (Stimpson): Figures 1, 2.—Interiors of the dorsal and ventral valve, x2, showing the characteristic septa, one in the dorsal, two in the ventral valve, hypotype USNM 111038a. *Locality*: Fort Macon, North Carolina.

Platidia davidsoni (Deslongchamps): Figures 3-6.—3, 4, Ventral and side views, x4, of a complete specimen showing the large foramen shared by the dorsal valve, hypotype USNM 550672a; 5, Ventral valve of *Ecnomiosa* with attached *P. davidsoni*, x1, hypotype USNM 550762b; 6, Ventral valve of the preceding specimen enlarged, x6, to show the papillose ornament. *Locality*: OREGON Station 4570 (number 9 on map).

Cryptopora rectimarginata Cooper: Figures 7-14.—7-11, Posterior, anterior, ventral, dorsal and side views, x6, of a complete individual, hypotype USNM 323891; 12, dorsal view, x8, of the ventral valve interior of the preceding specimen showing elevated deltidial plates and teeth; 13, 14, Laterally tilted and ventral views, x8, of the interior of the dorsal valve of the preceding specimen showing the maniculifer crura and the short, strongly elevated median septum. *Locality*: ALBATROSS Station 2400 (number 21 on map).

Tichosina cubensis (Pourtalès): Figures 15-22.—15-19, Anterior, side, posterior, ventral and dorsal views, x1, of an unusually large specimen, hypotype USNM 550755. *Locality*: OREGON Station 1189 (number 3 on map).—20-22, Anterior, side and dorsal, views, x1, of a young specimen showing rectimarginate anterior commissure, hypotype USNM 550761. *Locality*: West-southwest of Dry Tortugas at 213 meters (R. Cooper collection) (DT on map).

194

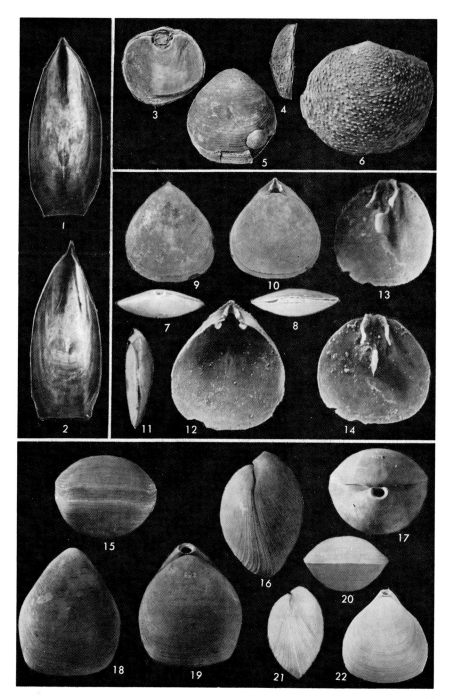

PLATE 28

Crania and *Terebratulina*

Crania aff. *C. pourtalesi* Dall: Figures 1-3.—1, 2, Exterior, x1, x2, of the dorsal valve, figured specimen USNM 61225; 3, Interior of the preceding specimen, x3, showing median ridge and muscle scars. *Locality*: Campeche Shelf (Bank), Mexico.

Terebratulina cailleti Crosse: Figures 4-27.—4, Two individuals in life position attached to a small sponge, x2, hypotype USNM 550756f; 5, 6, Another specimen, x1, x2, attached to a fragment of shell, hypotype USNM 550756g; 7, Another specimen attached to a piece of shell, x1, hypotype USNM 550756h; 8, Dorsal view of a complete specimen, x1, hypotype USNM 550756b; 9-13, Anterior, posterior, ventral, dorsal and side views of the preceding specimen, x2; 14-18, Dorsal, anterior, posterior, ventral and side views of a large specimen, x2, showing beaded costellae at posterior, hypotype USNM 550756a; 19-23, Anterior, ventral, dorsal, posterior and side views of another individual having beaded costellae, x2, hypotype USNM 550756d; 24, Interior of the dorsal valve of the preceding specimen, showing the loop and a small transverse cardinal process, x3; 25, 26, Interior of the ventral valve of the preceding specimen, x3, x6, showing large spicules in parts of the adhering mantle; 27, Interior of another dorsal valve, x3, showing the strongly spiculate lophophore, hypotype USNM 550756e. *Locality*: OREGON Station 955 (number 16 on map).

197

PLATE 29

Chlidonophora and *Dallina*

Chlidonophora incerta (Davidson): Figures 1-13.—1-5, Ventral, posterior, anterior, side and dorsal views, x3, of a young specimen, hypotype USNM 550757b; 6-10, Posterior, anterior, side, dorsal and ventral views, x3, of a large specimen, hypotype USNM 550757a, unnumbered, same, x1; 11, 12, Interior of the ventral and dorsal valves, x4, showing the large teeth, rudimentary deltidial plates of the ventral valve and the strong socket ridges and loop of the dorsal valve, hypotype USNM 550757c (contrast this loop with that of *Terebratulina* on the preceding plate); 13, Dorsal view of another large specimen, x4, showing the frayed pedicle characteristic of the species, hypotype USNM 550602a. *Locality:* OREGON Station 2574 (number 12 on map).

Dallina floridana (Pourtalès): Figures 14-24.—14-18, Dorsal, posterior, ventral, side and anterior views, x1, of an adult showing its complicated anterior folding, hypotype USNM 550758a; 19, Dorsal view of the preceding specimen, x1½, showing large foramen; 20, 21, Posterior and ventral views, x1½, of the dorsal valve interior showing the wide-ribboned loop, hypotype USNM 550758b. *Locality:* OREGON Station 4574 (number 8 on map).—22, Ventral view, x3, of the interior of a dorsal valve showing the median septum, concave inner and outer hinge plates and crural bases, hypotype USNM 110861b. *Locality:* Straits of Florida—23, Dorsal view, x10, of an immature specimen showing alate deltidial plates and large deltoid foramen, hypotype USNM 432772; 24, Interior of the dorsal valve of the preceding specimen, x10, showing the initial ring on the end of the septum, the descending branches not yet welded to the septum at the base of the ring and lack of inner hinge plates. *Locality:* ALBATROSS Station 2388 (number 14 on map).

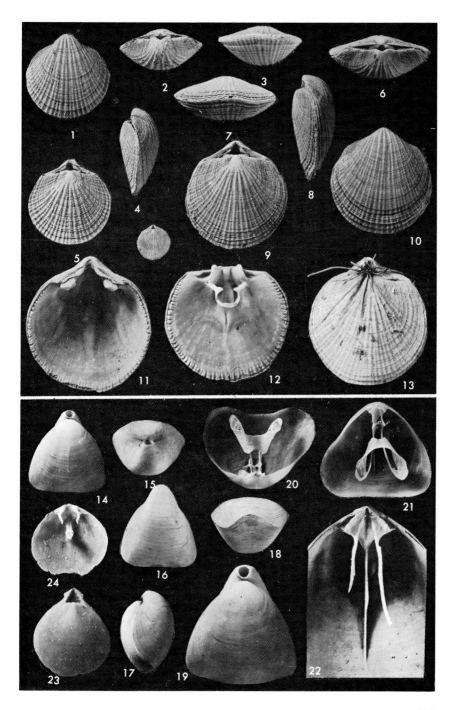

199

PLATE 30

Tichosina

Tichosina ovata Cooper, n. sp.: Figures 1-20.—1-5, Ventral, side, posterior, anterior and side views of an adult, x1, paratype USNM 549434c; 6-10, Posterior, anterior, ventral, side and dorsal views, x1, of a large individual, paratype USNM 549434b; 11, Dorsal view, x1, of a specimen smaller than the preceding, paratype 549434d; 12, rubber cast of the interior of a ventral valve, x1, showing the muscle and pallial marks, made from paratype USNM 549434-1; 13-17, Posterior, anterior, side, dorsal and ventral views of a large specimen, x1, holotype USNM 549434a; 18, Posterior of the ventral valve showing symphytium and teeth, x2, paratype 549434-1 (same specimen from which rubber cast of figure 12 was prepared); 19, 20, Posterior of two dorsal valves, x2, showing the loop and strongly elevated crural bases forming the margin of the outer hinge plates, paratypes USNM 549434m, n. *Locality*: OREGON Station 1408 (number 13 on map).

201

PLATE 31

Argyrotheca and *Stenosarina*

Argyrotheca sp. 4: Figures 1-12.—1-5, Side, dorsal, posterior, anterior and ventral views of a complete specimen, x6, figured specimen USNM 550759a; 6-10, Ventral, anterior, posterior, dorsal and side views, x6, of another adult, figured specimen USNM 550759b; 11, 12. Interior of the dorsal and ventral valves, x8, of the same specimen, figured specimen USNM 550759c. *Locality*: SILVER BAY Station 961 (number 4 on map).

Stenosarina oregonae Cooper, n. sp.: Figures 13-18.—13, Rubber cast of the interior of the ventral valve, x1, showing muscle scars, prepared from paratype USNM 550595b; 14-18, Side, posterior, dorsal, anterior and ventral views, x1, holotype USNM 550595a. *Locality*: OREGON Station 4574 (number 8 on map).

Stenosarina parva Cooper, n. sp.: Figures 19-25.—19, Dorsal view of the holotype, x1, USNM 550596; 20-23, Dorsal, anterior, posterior and side views, x2, of the holotype; 24, 25, Interior of the ventral and dorsal valves, x2, showing the narrow beak region and long slender loop of the holotype. *Locality*: Johnson-Smithsonian Expedition Station 43 (not on map).

Stenosarina angustata Cooper, n. sp.: Figures 26-33.—26-30, Anterior, dorsal, ventral, posterior and side views, x1, of the holotype USNM 550594; 31, side view of the holotype, x1 ½, showing the truncated beak; 32, 33, Interior of the dorsal and ventral valves of the holotype, x2, showing the characteristic narrow loop. *Locality:* OREGON II Station 11133. (number 11 on map).

202

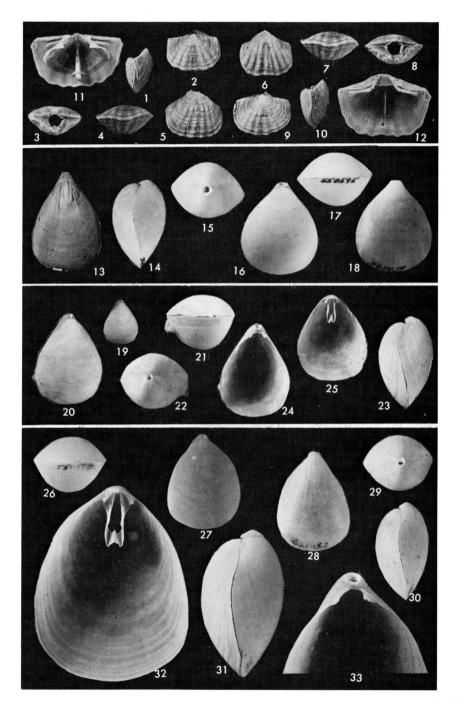

PLATE 32

Argyrotheca and *Stenosarina*

Argyrotheca hewatti Cooper, n. sp.: Figures 1-15.—1-5, Ventral, anterior, dorsal, posterior and side views, x4, of a complete specimen with posterior partially resorbed by growth pressure, holotype USNM 550739d; 6, 7, Ventral and dorsal exterior of a more normal specimen than the preceding, x4, paratype USNM 550739b; 8, 9, Interior of the preceding dorsal valve seen in ventral view, x6, and side view, x4, showing interior thickenings; 10, dorsal view of a complete specimen, x5, paratype USNM 550739a (valves parted during photography of the specimen); 11, Interior of the ventral valve, x5, showing thickened median septum, paratype USNM 550739e; 12, 13, Interior of the dorsal and ventral valves, x5, of another specimen showing the ventral median septum and the great thickening of the inside of the dorsal valve, paratype USNM 550739c; 14, 15, Interior of the dorsal valve, seen in ventral view, x6, and posteriorly tilted, x5, to show the great thickening of the muscle region to form an irregular platform, paratype USNM 550739f. *Locality*: 150 miles southwest of Sabine Pass, Texas (number 25 on map).

Stenosarina nitens Cooper, n. sp.: Figures 16-21.—16-20, Anterior, posterior, side, ventral and dorsal views, x1, of a young specimen, paratype USNM 550597. 21, Interior, x2, of the dorsal valve of the same paratype showing the long slender, anteriorly tapering loop. *Locality*: SILVER BAY Station 5181 (not on map).

Argyrotheca barrettiana (Davidson): Figures 22-32.—22-26, Dorsal, anterior, side, ventral and posterior views, x4, of a complete specimen, hypotype USNM 64229a; 27, 28, Interior and exterior of a large dorsal valve, x4, showing median septum and muscle scars but with loop broken away, hypotype USNM 64229b; 29, Side view of a small dorsal valve, x4, showing non-serrate median septum, hypotype USNM 64229c; 30, 31, Interior and exterior of a ventral valve, x4, showing short median septum, hypotype USNM 64229d (note borings in shell and consequent swellings on the inside of the shell); 32, Interior of a large and somewhat thickened dorsal valve, x4, showing muscle scars but with loop broken away, hypotype USNM 64229e. *Locality*: BLAKE Station 45 (number 18 on map).

205

PLATE 33

Stenosarina, Tichosina, Pelagodiscus, Platidia and *Dallithyris*

Stenosarina nitens Cooper, n. sp.: Figures 1-6.—1-5, Posterior, side, anterior, ventral and dorsal views of the holotype, x1, showing curved lateral commissure, rectimarginate anterior commissure and swollen dorsal umbo, USNM 550763; 6, Interior of the dorsal and ventral valves of the holotype, x3, showing small teeth, symphytium and long tapering loop. *Locality*: OREGON Station 5927 (not on map).

Stenosarina cf. *S. angustata* Cooper, n. sp.: Figures 7-11.—Anterior, posterior, side, dorsal and ventral views, x1, of a young specimen, figured specimen USNM 550766. *Locality*: OREGON Station 4939 (not on map).

Tichosina elongata Cooper: Figures 12-14.—Anterior, side and dorsal views, x1, of an unusually narrow species, figured specimen USNM 64249. *Locality*: BLAKE Station 100 (not on map).

Platidia anomioides (Scacchi and Philippi): Figures 15-17.—Ventral, side and dorsal views, x4, of a large complete specimen showing the concave dorsal valve and the large foramen at its apex, hypotype USNM 550760. *Locality*: ALAMINOS Station 9 (number 15 on map).

Dallithyris murrayi Muir-Wood: Figure 18.—Interior of the dorsal valve of *Dallithyris murrayi,* Muir-Wood type species of the genus, x2, showing the loop, hypotype USNM 550332. Introduced for comparison with the loops of *Tichosina* and *Stenosarina*. *Locality*: Maldive Islands, Indian Ocean.

Platidia clepsydra Cooper: Figures 19-27.—19, 20, Side and dorsal views, x10, of paratype USNM 550748b; 21-23, Side, ventral and dorsal views, x10, of the holotype USNM 550748a; 24, ventral view of a small specimen, x10, attached to a fragment of coralline algae, paratype USNM 550748f; 25, posterior of the ventral valve, x20, showing long dental plates, marginal deltidial plates and thick teeth, paratype USNM 550750e; 26, interior of the dorsal valve, slightly tilted laterally, x20, to show the naked loop, paratype USNM 550750d; 27, interior of another dorsal valve, x20, showing the lophophore, paratype USNM 550749a. *Locality*: Hourglass Station M.

206

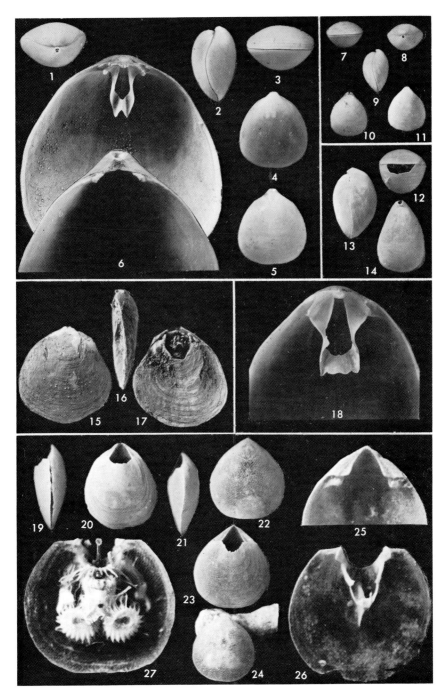

PLATE 34

Ecnomiosa

Ecnomiosa gerda Cooper, n. sp.: Figures 1-15.—1-5, Dorsal, side, posterior, anterior and ventral views, x1, of the holotype USNM 550510a; 6, 7, Ventral and partial anterior views, x1, of the interior of the dorsal valve of the holotype; 8, 9, Same views as the preceding, x2, showing the posterior ring of the loop attached to the short median septum, the small spines on the anterior part of the loop and only small nubs showing the position of the lateral bars; 10, 11, Anterior and dorsal views of the interior of the ventral valve tilted to show the dental plates, x1, holotype USNM 550510a; 12, Posterior of the ventral valve of the holotype, x2; 13, Interior of another ventral valve tilted to show the dental plates, x1, paratype USNM 550510b; 14, 15, Interior of the dorsal valve, x1, x2, showing hinge plate, short median septum, no cardinal process and pallial trunks, paratype USNM 550510c. *Locality*: G114.

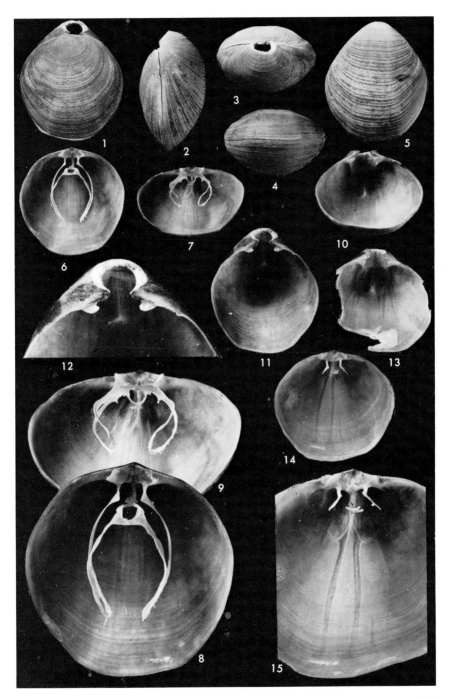

209

PLATE 35

Ecnomiosa

Ecnomiosa gerda Cooper, n. sp.: Figures 1-14.—1, 2, Dorsal and side views of an un-usually large specimen, x1, paratype USNM 550606a; 8, Interior of an immature dor-sal valve, ventral view, x1, showing campagiform loop, paratype USNM 550589a; 9, Same specimen as preceding tilted posteriorly to show anterior of loop, x3; 10-12, Ven-tral, anteriorly tilted and partial side views of the preceding immature dorsal valve, x6, showing campagiform loop; 13, 14, Dorsal valve interior tilted posteriorly and in ven-tral view, x2, showing nearly adult loop still retaining lateral ribbons attaching the de-scending branches to the median septum (terebrataliform stage), paratype USNM 550600. *Locality*: OREGON II Station 10962.—3, 4, Latex impressions of the interior of the dorsal and ventral valves of a paratype USNM 550590b, x1, showing muscle scars and pallial sinuses; 5-7, Ventral view, x1, and ventral and laterally tilted views, x2, of a dorsal valve showing nearly adult loop, paratype USNM 550590a. The lateral bars ty-ing the descending branches are visible on each side, the left one (reader's) complete but the right one partially resorbed. *Locality*: OREGON Station 4570.

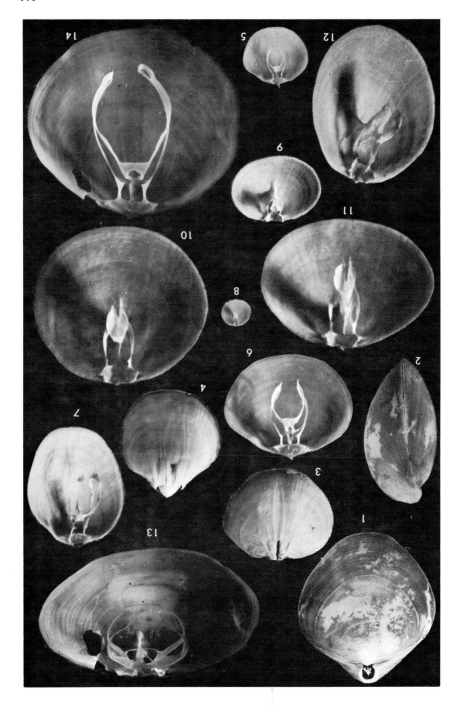